Nutritional Disorder in Fruit Crops
Diagnosis and Management

Nutritional Disorder in Fruit Crops
Diagnosis and Management

M. Prakash
Professor
Faculty of Agriculture
Annamalai University
Annamalainagar – 608 002
Tamil Nadu

K. Balakrishnan
Professor of Crop Physiology
Agricultural College & Research Institute
Tamil Nadu Agricultural University
Madurai – 625 104
Tamil Nadu

A. Rathinasamy
Professor of Soil Science & Agricultural Chemistry
Horticultural College & Research Institute
Tamil Nadu Agricultural University
Periyakulam – 626 504
Tamil Nadu

NEW INDIA PUBLISHING AGENCY
New Delhi – 110 034

NEW INDIA PUBLISHING AGENCY
101, Vikas Surya Plaza, CU Block, LSC Market
Pitam Pura, New Delhi 110 034, India
Phone: + 91 (11)27 34 17 17 Fax: + 91(11) 27 34 16 16
Email: info@nipabooks.com
Web: www.nipabooks.com

Feedback at feedbacks@nipabooks.com

ISBN: 978-93-81450-95-6

Composed, Designed and Printed in India

Preface

In recent years, the concept of nutrient management has undergone many changes. With the tremendous advances made in the understanding of the role of various nutrients, greater emphasis has now been given to secondary (Ca, Mg, S) and micro nutrients (Zn, Fe, Cu, Mn, Mo, B, Cl) besides N, P and K in nutrient management of crop plants. Long and continuous exploitation of soil fertility due to advent of high yielding varieties has resulted in the occurrence of nutrient deficiencies in general and micro nutrient deficiencies in particular. Diagnosing the nutrient deficiency is the main prerequisite for taking corrective measures so as to realize the maximum yield potential in fruit crops.

Diagnosis involves careful observation of crop with a thorough knowledge of crop behaviour as well as a complete understanding of functions of nutrients and their deficiency symptoms. Diagnostic techniques vary from crop to crop depending upon the field condition in which the deficiency occurs. Indian agriculture has entered into an era of multiple nutrient deficiencies. Hence, knowledge on these aspects should be imparted to the students of agriculture to face the forth-coming challenges in nutrient management.

Realizing the gravity of emerging problem of nutrient deficiencies and to fulfill an immediate need to tackle them more efficiently, a guide on diagnosis of nutritional disorders and their corrections in crop plants has been prepared. All the new diagnostic techniques have been discussed in a simple language. The book has been designed in such a way as to improve the knowledge on diagnosis of deficiencies of mineral elements essential for normal plant growth, and of the methods by which such deficiencies may most effectively be remedied. The main feature of this book is the detailed description of the various visual deficiency symptoms exhibited by the fruit crops.

The book has been written primarily for the use of the students of Horticulture and Agriculture to help update their knowledge on nutritional and physiological disorders of fruit crops. It is also felt that the book will provide a suitable basis for those engaged in the profession of agriculture like extension workers and progressive farmers. We hope that this book will go a long way to help in increased fruit production by proper diagnosis and suitable correction of nutritional disorders in fruit crops.

Authors

Contents

CHAPTER 1

Concept of Plant Nutrition

1.1. Introduction

The green revolution has led our nation from the state of deficiency to the state of self-sufficiency in food production. The grain production has increased substantially from around 50 million tonnes to about 220 million tonnes per annum. As per the latest estimates, annual fruit production is 44 million tonnes from an area of 3.72million hectares.

Depletion of nutrients, from the soil is steadily increasing due to the adoption of modem agricultural technology through intensive and extensive cultivation ofhigh yielding varieties with chemically pure, micronutrients free high analysis fertilizers. So in recent years, the potentiality of many crops could not be exploited owing to the widespread nutrient deficiency observed in many parts of our country. With nutrient removals by crops exceeding the additions, there exists a perpetual negative balance. Consequently, nutrient deficiencies have multiplied in Indian soils and crops.

1.2. Essentiality of plant nutrients

Plants require food for their growth and development. The food of the plant consists of certain chemical elements which is referred to as plant nutrients. All green plants are capable of synthesising carbohydrates in the presence of light. In 1840, Liebig originated the concept of elements as plant nutrients by his revolutionary theory of mineral nutrition.

Plant absorb more than 90 elements from the soil. Of these 25 elements such as Carbon (C), Hydrogen (H), Oxygen (O), Nitrogen (N), Phosphorus (P),

Potassium (K), Calcium (Ca), Magnesium (Mg), Sulphur (S), Manganese (Mn), Molybdenum (Mo), Copper (Cu), Boron (B), Zinc (Zn), Iron (Fe), Chlorine (Cl), Sodium (Na), Silicon (Si), Aluminum (Al), Chromium (Cr), Arsenic (As), Bromine (Br), Gallium (Ga), Lanthanum (La), Lithium (Li), Lead (Pb), Vanadium (V), Yttrium (Y), Zirconium (Zr) and Selenium (Se) are considered for nutrition of plants. Among these, only 16 elements *viz*, C, H, O, N, P, K, Ca, Mg, S, Zn, Cu, Fe, Mn, Mo, B and Cl are considered as essential elements based on the criteria of essentiality laid dune by Arnon (1954).

1.2.1. The criteria of essentiality of elements

1. A deficiency of the element makes it impossible for plant to complete the vegetative or reproductive stage.
2. The deficiency is specific to that particular element and can be corrected by supplying only that particular element.
3. The element should play a direct role in the metabolism of the plant.

1.3. Clssification

1.3.1. Based on the nutrient contents

These 16 elements are further classified in to two groups *viz.*, 1. Macroelements or Majornutrients and 2. Micro elements or Micronutrients. Macroelements or Macronutrients consist of C, H, O, N, P, K, Ca Mg and S. These elements are required by plant in relatively large quantities and need to be found in the plant in quantities greater than 0.02%. Of these C, H and O are mostly absorbed from air and water in the form of CO_2 and H_2O. They are available in abundance in the environment and hence there is no need to supplement by the addition of fertilizer. Out of these, N, P and K are known as primary nutrients and Ca, Mg and S are called as secondary elements or secondary nutrients. These six elements *viz.*, N, P, K, Ca, Mg and S are available in limited quantities in soil medium and hence they should be adequately supplied as fertilizer for growth and development of plants.

The other 7 elements *viz.*, Zn, Cu, and Fe, Mn, B, Mo, and Cl called as microelements or micronutrients are needed relatively in small amounts for the plant growth. Sometime these are also called as trace elements. These are found in the plants in quantities between 0.0002 and 0.02%. These nutrient ions in the soils are continuously exhausted and hence deficiencies are not uncommon. Hence, these elements are to be supplied to the soil as fertilizers to increase their availability or to correct their deficiency under defict conditions.

Besides the above 16 elements, the other elements, Cobalt (Co), Nickel (Ni), Silicon (Si), Sodium (Na), Vanadium (V), Rubidium (Rb), Strontium (Sr), Chromium (Cr) and Arsenic (Ar) are considered as beneficial elements as they have been shown to stimulate the growth of certain plants. Hence, from the classification of elements, it is evident that the three basic elements C, O and H which plant takes from the environment without being necessarily to be added as fertilizers are not included in the following text.

1.3.2. Based on chemical nature and mobility of ions

The nutrient ions found in plant tissues are classified into different categories according to their chemical nature, mobility of ions and their role in synthesis of organic compounds as follows:

I. Chemical nature

a. Metals : K, Ca, Mg, Fe, Zn, Cu, Mn, etc.

b. Non metals : O, C, H, N, P, S, Cl, etc.

c. Cations : K---$^{+}$, Ca^{++}, Mg^{++}, Fe^{++}, Zn^{++}, Cu^{++}, Mn^{++}

d. Anions : NO_3^-, $H_2PO_4^-$, SO_4^{--}

II. Mobility in plants

a. Highly mobile : NO_3^-, Na^+, Mg^{++}, $H_2PO_4^-$, K^+, and Cl^-

b. Less mobile : S_4^{--}, Fe^{++}, Mn^{++}, Zn^{++}, and MoO_4^{++}

c. Immobile : Ca^{++} and $H_2BO_3^-$

CHAPTER 2

Principles of Diagnosis

Diagnosis is the art of recognizing diseases or deficiency symptoms manifested from their symptoms, a scientific determination of a disorder based on critical scrutiny.

2.1 Diagnosis and prognosis

Diagnosis is the estimation of nutritional status of a plant during sampling, while prognosis is the prediction of possibility of deficiency impairing plant growth at stages in the growth cycle after the sample is taken. Diagnosis is based on comparison of nutrient concentration in plant parts (mainly leaves) with standard values related to physiological requirements of the plants. In diagnosis, mainly plant analysis is used. Prognosis is also based on critical concentration of same plant parts. This sometimes has led to confusion between these two concepts, which have been used with the same meaning. In contrast, to diagnosis, prognosis may be based on their leaf or soil analysis. The main difference between prognosis and diagnosis lies in the time between sampling and appearance of the influence on plant growth.

2.2 Diagnostic tools

Diagnosis require experience and knowledge. These are many pitfalls. To a good job, the diagnosis should begin with observations made in the field. This often requires frequent visit to the same field. The diagnostician should have

- A spade to look at roots and environment
- A knife – to look at intenal tissues.

- A soil tube- to look for compaction and profile
- A hand lense – to look for disaeses or insects
- A note book – to record history and symptoms
- An "open mind" – to avoid pre-conception and bias.

2.3 Steps involved in diagnosis

The diagnosis of nutrient deficiency involves the followings six steps.

1. The realization may be something is wrong with the crop and that the problem might be avoidable.
2. The observation in detail of all the abnormalities exhibited by the crop, and how these are distributed on the plant and across the field.
3. The identification of possible cause by reference to the colour atlas. If any picture resemble the symptom found, one should not jump to conclusion but examine all pictures of the crop in question.
4. The elimination of unlikely explanations, using the informations about the effect of soil type, pH and species susceptibility.
5. If doubt remains, it will be necessary to check the diagnosis by means of leaf or whole plant analysis.
6. Definite proof of the correctness of the diagnosis is only really possible by successful application of curative treatments and comparison with untreated plants grown alongside the treated plant and cultivated exactly in the same way.

2.4 Diagnosing nutritional disorders

Plants speak through distress signals. The message may tell us there is simply a shortage of water or there is too much water or the signal may tell of a disease caused by an organism such as a virus or bacteria or damage due to insects nematode or rodent. There may be injuries from frost, lighting, pesticide, or mechanical equipments. Similarly if the essential nutrients are not in required concentrations, the plant tell us through its deficiency symptoms.

These symptoms are nutrient specific and show different patterns in crops for different essential nutrients. One has to observe carefully and let the plant tell what it can by its visual symptoms. It is a good tool to detect deficiencies of nutrients in the field. However, it requires learning to identify nutrient deficiencies. In applying this technique, one must develop diagnostic proficiency through practice and close observations.

These symptoms include, marginal and interveinal chlorosis of leaves, marginal leaf scorch, necrotic areas or spots in leaf or fruit tissue, bark necrosis, death of meristems and shouts, high pigmentation of leaves, stems and fruits, leaf rosetting; failure of the leaf lamina to develop and partial or complete dwarfing of plants.

2.5 Development of deficiency

The expression of deficiency symptom is due to the altered physiology caused by deficiency of an element. Deficiency or absence of any one of the essential nutrient element leads to disruption in normal morphology and physiology of the plant. It is logical to believe that abnormalities do not appear suddenly, but there must be some period between the beginning of subtle change at the molecular level and manifestation in gross from on the plant morphology. In other words, these changes affected through nutrition imbalance must begin early from the formation of cell components leading to the tissue and organ development. Therefore, considerable damage has already occurred to the plant by the time the abnormalities are macroscopically visible in the form of what is commonly called visual deficiency symptoms.

Some of the common symptoms for all the nutrients are given below

Bronzing : Development of bronze/copper colour on the tissue

Chlorosis : Loss of chlorophyll content results in paleness / yellow colour

Decline : On set of general weakness as indicated by loss of vigour, resulting in poor growth and reduced yield.

Firing : Burning of tissue from top to bottom accompanied with dark brown or reddish brown Colour

Lesions : A localized wound of the tissue accompanied with loss of normal Colour.

Necrosis : Death of tissue

Scorching : Burning of tissue accompanied with light brown colour.

Gummosis : Oozing out of the cell sap in the form of gum.

Distortion of Leaves : Irregular shaping of leaves viz, cupping, twisting, hooking, curling, laddering etc.

Mottling : Leaf surfaces marked with coloured spots due to anthocyanin pigmentation.

Rosetting : Clustering of leaves due to reduced leaf size and shorter internodal distance.

Die-back : Collapse of growing tips – affecting the youngest shoot

CHAPTER 3

Diagnostic Techniques

Several diagnostic techniques are employed to asses the nutrient status of the crop. These are helpful in determining specific or multinutrient stresses and quantity of nutrients needed to eliminate the stress. Some of the techniques commonly used are:

1. Visual deficiency symptoms
2. Plant analysis
3. Rapid tissue testing

3.1. Use of Visual deficiency symptoms

Nutrient deficiency symptoms can be used to determine the nutrient needs of crops, especially when they are specific for a particular nutrient. Yellowing along the midrib and eventual drying of lower leaves is a specific indication of N deficiency; whereas chlorosis and necrosis on the outer edge of lower leaves specifically indicate K deficiency.

Many symptoms can be caused by any one of several nutrients or by other conditions. Yellowing and interveinal chlorosis of young, newer leaves could be due to S deficiency or it could indicate the deficiency of one of the metal micronutrients viz., Fe, Zn, Mn, or Cu. A knowledge of soil pH and general soil conditions may be needed to determine whether the yellowing results from lack of S or any one of the micronutrients. Sulphur deficiency symptoms seldom appear on high-pH soils; whereas deficiencies of Fe, Zn, Mn, or Cu seldom occur on low-pH soils.

Typical deficiency symptoms can also be caused by many other conditions. Certain herbicides, diseases, and insects can cause chlorosis in plants. Water-logged or droughty soils and mechanical or wind damage can often create problems that mimic deficiencies.

Deficiency symptoms should be used only as a guideline. Symptoms, plant and soil analyses together with a general knowledge of crop requirement and the chemistry of the soils should all be used in determining crop nutrient needs.

3.1.1 Distribution of visible deficiency symptoms

One of the criteria for mineral deficiency diagnosis is the mobility of nutrient in phloem vessels of plants. Based on the phloem mobility, essential elements are classified in to three groups as mobile, less mobile and immobile. The distribution of visible deficiency symptoms and remobilization of mineral nutrients vary among plant parts. The remobilization of mineral nutrients in plant takes place form older leaves to younger leaves.

3.1.2 Remobilization of nutrient and distribution of deficiency symptoms.

Distribution of deficiency symptom		Remobilization of nutrients.
Nutrient	Plant part	
N, K, Mg and P S	Older leaves	Excellent
Fe, Zn, Cu and Mo	Young leaves	Low
Ca and B	Young leaves	Very low
	Growing tips	Nil

Based on phloem mobility, remobilization of mineral nutrients and distribution of visual deficiency symptoms, quick diagnosis can be made. In order to achieve this objective, the following flow chart can be used.

3.1.3 Distribution of nutrient deficiency symptoms in crops

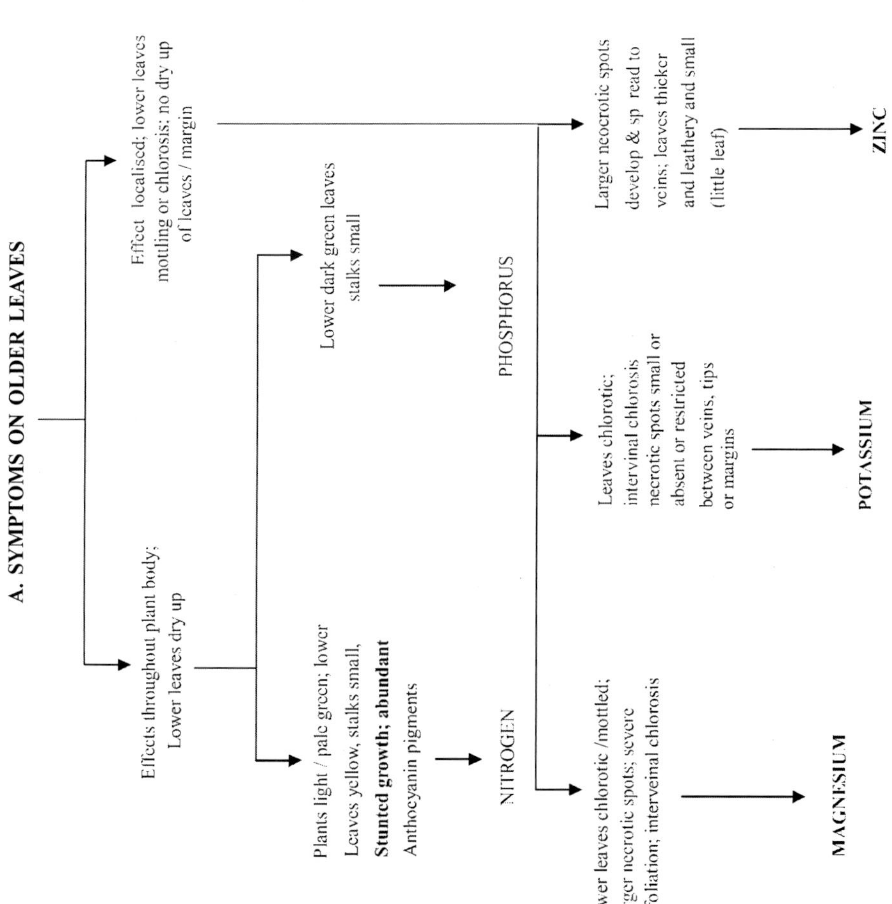

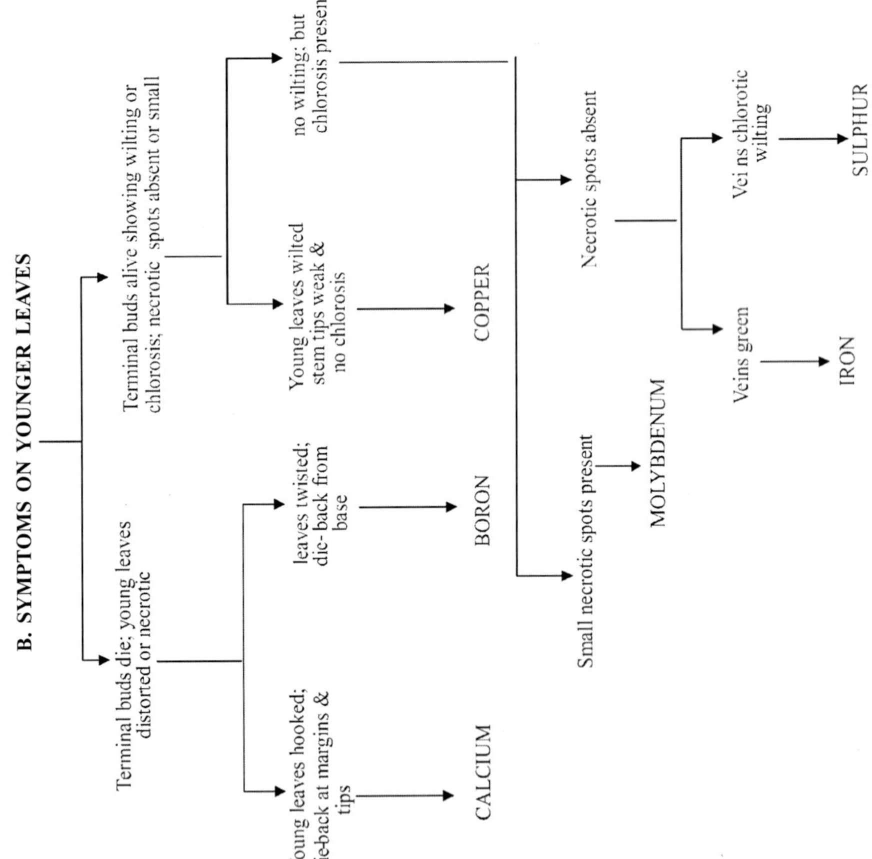
B. SYMPTOMS ON YOUNGER LEAVES
Terminal buds die; young leaves distorted or necrotic
Terminal buds alive showing wilting or chlorosis; necrotic spots absent or small
Young leaves hooked; Die-back at margins & tips
leaves twisted; die-back from base
Young leaves wilted stem tips weak & no chlorosis
no wilting; but chlorosis present
CALCIUM
BORON
COPPER
Small necrotic spots present
Necrotic spots absent
MOLYBDENUM
Veins green
Vei ns chlorotic wilting
IRON
SULPHUR

3.2. Plant analysis

The concentration of cations and anions in the plant tissues indicate the accessibility of the concerned elements from the soil to the plant. To some extent, there exists a relationship between the concentration of an element (except potassium) in the plant and the total biomass of the plant. A correct balance of nutrient ions in the plant tissue is closely associated with the maximum yield.

Plant analysis have been used as a technique for many years not only to diagnose but also to monitor the nutritional status of the crops. It is also helpful to solve so many other field related problems.

1. To diagnose or to conform a diagnosis based on visible symptoms which may be due to deficiency or toxicity.
2. To identify Sub clinical disorders (Hidden hunger) in the absence of visible symptoms.
3. To assess the nutritional status of the crops which may throw lights on the imbalance of nutrient ions or interactions among nutrient ions.
4. To find out the effect of applied nutrients on plant tissue.
5. For understanding the internal functioning of nutrients.
6. To make an assessment of removal and residues of the key elements and thus helping in rational recommendation of fertilizers for getting higher yield.

Plant analysis as a guide to the fertilization of crops is based on the concept that "what is in the plant is related to growth". Nutrient concentrations within the plant commonly decrease with growth, and if the decrease is large enough, the growth rate becomes lesser than that of plants at a higher nutrient concentration (Fig.1).

The critical nutrient concentration is that the concentration at which the plant's growth rate begins to decrease compared to plants having a higher nutrient concentration (assuming all other growth factors are present in ample supply). It follows that the longer a plant remains below the critical concentration and the earlier in the growing season the deficiency occurs, the lesser the growth or yield and the greater the likelihood of a response to fertilization.

In essence, the crop through plant analysis informs the farmer that the power of the soil to supply a nutrient has not kept with the nutrient requirements of

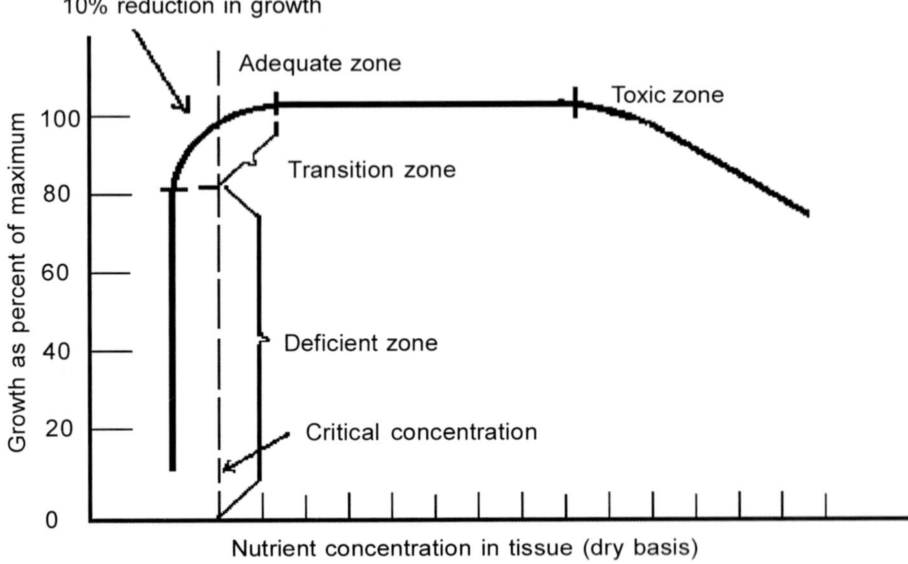

Fig.1: Growth of a crop related to the concentration of a nutrient in a tissue.

the crop. Consequently, if the crop is to grow satisfactorily, supplies of the nutrient must be increased either directly by fertilization or indirectly by enhancing the nutrient- supplying power of the soil through improved cultural practices, such as crop rotation, improved soil structure, better varieties, improved irrigation, better drainage, adequate plant spacing, and better weed control.

3.2.1 The critical concentration

It is essential to select a plant part for chemical analysis that reflects the nutritional status of the plant for most or all of the growing seasons. For best results, the plant part selected should remains relatively constant in nutrient concentration in the zone of deficiency and then increase rapidly in concentration as the plant becomes well supplied with nutrients. The transition zone from deficiency to adequacy should also be relatively narrow; just as in using an indicator in the titration of an acid with a base, the indicator changes color instantly at the end-point. Similarly, a plant sample taken any time during the growing season will tell us about the nutrient status of the crop simply by comparing the analytical results to the critical concentration; if our plant test is a sensitive one, as reflected by a sharp transition zone from deficiency to adequacy, then our answer will be a definitive yes or no. Nutrient excesses in terms of toxicity are not so clearly defined as nutrient deficiencies, but when plant analysis values are unusually high, the cause for the high values should be investigated.

3.2.2 Plant sampling

The successful application of technique depends upon the selection of the right plant part at correct stage of plant growth. The whole plant analysis is done for working out the total nutrient uptake. Analysis is done for whole plant or plant parts like leaf, petiole, etc. Thus, depending upon the purpose of analysis, plant-sampling needs to be planned. There are certain guidelines which are to be followed for while sampling are.

1. Basal, older and young terminals are avoided because older tissues lose some of the mobile nutrients and accumulate the immobile ones such as calcium; while young immature tissues are generally very rich in mobile elements
2. Recently matured leaf from non-fruiting or vegetative branches has been considered as the most suitable. These plant parts of known physiological age are sensitive to nutrient supply.
3. Samples from all the four directions *viz.*, West, North, South, East and West should be selected and pooled.
4. The size of the samples depends upon the leaf size *viz.*, 25 – 50, if leaves are small sized, 20 – 30 if leaves are average sized and 10 – 15 if leaves are from large sized.
5. Collect leaves from the trees fully exposed to sun. Similarly, select leaves which are exposed to sun and avoid leaves in shade conditions.
6. Take samples prior to irrigation or fertilization
7. A small area of crop (0.5 – 1.0 ha) judged as representative of the average crop condition in the field is to be selected.
8. Typical sampling instructions involve (a) selection of a uniform area of orchard (soil; cultivar, age) and within this area, select 20 trees along on X or Zig Zag passage through the orchard, (b) collect four leaves per tree, one from each of the north, south, East and west quarters of the tree (c) select leaves at shoulder height.

Table 1: Plant parts recommended to be sampled for analysis of fruit corps

Crop	Part to be sampled with stage/age
Almond	3rd leaf from top, at the beginning of bloom
Apple, Pear	Leaves from middle of terminal shoot growth, 8-12 weeks after full bloom, 2-4 weeks after formation of terminal buds in bearing trees
Blackberry	Latest matured leaf from non-tipped canes, 4-6 weeks after peak bloom.
Cherry	Fully expanded leaves, fruiting mid shoot current growth in July-August
Peach	Mid-shoot leaves, fruiting or non-fruiting spurs, mid-summer leaves, fruiting
Strawberry	Youngest, fully expanded matured leaf without petiole, at peak or harvest period
Plum	Leaves from middle of current seasons extension growth, in January-February
Banana	Petiole of 3rd open leaf from apex, 4 months after planting
Cashew	4th leaf from tip of matured branches, at beginning of flowering
Grapes	5th Petiole from base at bud differentiation to yield, and petiole opposite to bloom for quality
Citrus fruits	3-5 month old leaves from new flush, 1st leaf of the shoot , in June.
Guava	3rd pair of recently matured leaves, at bloom (August or December)
Mango	Leaf with petiole (4-7 month old) from middle of shoot
Papaya	6th petiole from apex, six months after planting
Pineapple	Middle 1/3 rd portion of white basal portion of 4th leaf from apex, at 4-6 months stage
Pomegranate	8th leaf from apex at bud differentiation, in April and August
Falsa	4th leaf from apex, one month after pruning
Ber	6th leaf from apex from secondary or tertiary shoot, 2 months after pruning.
Carambola	Middle pair of leaflets from median leaves located on terminal shoots when fruits attains ½ maturity.
Litchi	Collect leaves from middle of latest mature flush of growth during flowering
Mangosteen	Select un branched, non fruiting terminals.
Passion fruit	Collect the leaves just beyond the last open flower
Sapota	Collect leaves during middle of the main blossoming period.
Date Palm	17th fronds, counting the last fully developed frond as number one
Jack fruit	7th or 8th leaf from the apex of terminal shoot.
Bread fruit	Select 5th or 6th leaf from the fruit
Fig	Collect leaves 10-14 weeks after the appearance of flowers at basal portion of new growth
Custard apple	Take leaves 12-14 weeks after the appearance of new growth from middle of the terminal shoot growth.
Avocado	4-5 months old leaves from non-fruiting shoots.
Loquat	Select leaves from middle of the shoots developed from the spring.
Tamarind	Collect leaves from middle of the growth produced by current terminal flush.

3.2.3 Sample handling and preparation

Each step in sample preparation may contribute errors which may have a cumulative effect on the final result. Hence, standard procedures should be employed to minimize or prevent such errors which are finished below.

1. Field sampling should be undertaken early in the week.

2. The collected samples should be placed in labelled open paper bags and immediately placed in cooled containers (5°C) for transport from the field to the laboratory within 24 hours of sampling.
3. If roots are to be included, uproot the whole plant carefully from wet soil, retaining even the fine/active roots. Gently dip the plant roots in water several times to remove adhering soil as far as possible.
4. Wash the plant parts with water several times.
5. Wash the samples with about 0.2% detergent solution thoroughly to remove the waxy/greasy coating on the leaf surface, which is often present.
6. Wash with 0.1N HCl followed by thorough washing with plenty of water. Give final rinse with distilled water.
7. Rinse with double distilled water, specifically if micronutrient analysis is to be carried out.
8. Soak dry with good quality tissue paper.
9. Air dry the samples on a perfectly clean surface at room temperature for at least 2-3 days in a dust-free atmosphere away from any kind of contaminant.
10. Put the samples in an oven and dry at 60°C for 48h.
11. Grind the samples in an electric stainless steel mill using 0.5 mm sieve. Clean the cup and blades of the grinding mill before each sample.
12. Put back the samples in oven and dry again for few hours more for constant weight. Store in well-stoppered plastic or glass bottles or paper bags for analysis.

3.2.4 Laboratory analysis

The dried plant samples are analysed for N, P, K, Ca, Mg, S, Zn, Cu, Fe, Mn, B, and Mo. The N content of the plant samples was estimated by microkjeldahl method (Humphries 1956). For analysis of P, K secondary and micronutrients, a measured quantity of dried samples is digested with tri acid mixture of nitric acid, sulphuric acid and perchloric acid @ 9:3:1. This extract has been used for the estimation of Ca and Mg (Versanate Titration Method), phosphorus and potassium (Jackson, 1962), Zn, Cu, Fe and Mn (Atomic absorption spectro photometer) (Lindsay and Norvell, 1978). The sulphur content of the plant can be estimated by barium chromate colorimetric method (Palaskar et al 1981). Boron content is estimated by the method suggested by John et al. (1975).

3.2.5 Interpretation of plant analysis

Nutrient content in plant tissues may serve as an aid to soil testing in specific cases. Most of the plant species show a specific ratio of various nutrient ions present in them as an indication of the sufficiency of the concerned elements in available form in soil. The supply of a mineral element to a crop may said to be sufficient when the growth or development of the plant is not affected by increasing the supply, that is the plant has as much as it need. When the plant has less than it needs, so that an increase in supply of the element improves growth or development, it is said to be deficient of that element. Sometimes, at the other extreme, a plant has too much of an element supplied to it, and growth or development is impaired. The supply of the element may then be said to be excessive.

The critical level of nutrients is a well-known concept for diagnosing nutrient deficiency in plants (Table 2). It indicates the single point on the curve relating nutrient content to crop yield. This level or range may be considered as the "just sufficient" or " minimum" for the highest yield. Obviously, a level below it will result in decreased yield. An adequate range also follows above the critical level, which is considered as the goal to be achieved while making suitable recommendations. However, there could be even the upper critical level of nutrient in the end of adequate range in the case of some nutrients, which may also adversely affect the yield. Nutrient balance in plant can be assessed from the ratio of various nutrients. The well-established nutrient ratios which indicates the antagonism between them are N/S, K/Mg, K/Ca, Ca + Mg/ K and N/P.

Table 2: General guideline of nutrient content of crop plants

Nutrients	Critical level	Sufficient range	Ionic form	Toxicity level
Nitrogen (%)	<2.0	2.0 – 5.0	NH_4^+, NO_3^-	Non toxic
Phosphorus (%)	< 2.0	0.2 – 0.5	$H_2PO_4^-$, HPO_4^{2-}	Non toxic
Potassium (%)	< 1.0	1.0 – 0.5	K^+	Non toxic
Calcium (%)	< 0.1	0.1 –1.0	Ca^{++}	Non toxic
Magnesium (%)	< 0.1	0.1 – 0.4	Mg^{++}	Non toxic
Sulphur (%)	< 0.1	0.1 – 0.3	SO_4^{2-}	Non toxic
Iron (ppm)	< 50	50 –250	Fe^{2+}, Fe^{3+}	Non toxic
Zinc (ppm)	15 – 20	20 –100	Zn^{2+}	> 400
Manganese (ppm)	10 – 20	20 – 300	Mn^{2+}	> 300
Copper (ppm)	3 –5	5 – 20	Cu^{2+}	> 20
Boron (ppm)	< 10	10 –100	$H_2BO_3^-$, H_3BO_3	> 100
Molybdenum (ppm)	< 0.1	0.1 – 0.5	MoO_4^{2-}	> 0.5
Chlorine (%)	< 0.2	0.2 – 2.0	Cl^-	> 2.0
Silicon (%)	< 0.2	0.2 – 2.0	$Si(OH)_4^-$	Non toxic

3.3 Rapid tissue testing

Rapid tissue testing is a simple technique usually used to identify the nutrient deficiency symptoms. This method is used to determine the easily soluble nutrient ion in any part of the plant. Generally, lamina or petiole has been found most satisfactory.

This test is being done with little chemical and the result is known very quickly. The cell sap of the suspected plant or leaf is allowed to react with the specific reagent and the intensity of the colour is observed. On the basis of the intensity of the colour, three categories will be obtained (1) deficient, (2) slightly deficient and (3) sufficient.

This method can also be employed to identify not only the deficiency but also to find out the excess (toxicity). The reactions are not so significant in intermediate situations. Nevertheless, they are very useful in practice when they are properly used.

3.3.1 Collection of cell sap

1. Cell sap may be obtained from squeezing the leaves, veins or petioles. Colour reactions are often hindered by the green colour of the sap. Activated charcoal or centrifugation may be used in order to clarify the solution for analysis.
2. Sap can be collected from plants by piercing the xylem or cutting the base of the stem (tendrils, shoots), after which it is bound tightly with a rubber tube and the sap is collected over a 24-hour period.
3. Fresh sap also may be collected from the grape vine through guttation (weeping)
4. In some cases, the sap may be absorbed directly onto a filter paper.

3.3.2 Collection of sample

Samples for analysis must be representative. Recently matured leaves may be collected between 7.00 and 9.00 am. Leaves may be collected from every 20th and 30th plant for field crops and fruit tree crops, respectively. Recently matured leaves were harvested from the middle third of the plant, stem or branch.

The nutrient status of various elements at different stages is an important factor for crop productivity. The nutrient status of the plant at a particular stage should be analyzed so as to supply the deficient nutrient in proper quantity for good crop yield.

The proper diagnosis becomes necessary to find out the requirement of different nutrients for confirmation of different symptoms and also for assessing the nutrient status at a particular stage of the plant.

There are two types of plant analysis

1. Tissue testing
2. Whole plant analysis

Tissue testing is done usually with fresh leaves of the plant in the field itself; whereas the total analysis is performed in the laboratory. These plant analysis methods are based on the assumptions that the particular element is an indicator of the supply of that particular nutrient. The whole plant analysis methods involve elaborate equipment and a lot of chemicals and cannot be performed in the field itself. Tissue testing is also very rapid. The test is made with fresh plant saps and is very useful in quick diagnosis. In this test, the sap from the cell is tested for unassimilated N, P and K. Test for Fe, Ca and Mg are also used frequently in a variety of crops.

3.3.3 Advantages

1. Spot tissue testing can be done easily and rapidly without any equipment.
2. More number of samples can be tested within a few minutes in the field itself.

In general, it is necessary to test the specified part of the plant which will give the best indication of the nutritional status.

1. Nitrogen

Reagents: 1% diphenylamine in conc. sulphuric acid. (1 gm in 100 ml conc. sulphuric acid)

Small bits of leaf or petiole are taken in a petridish and a drop of 1% diphenylamine is added. The development of blue colour indicates the presence of nitrate-nitrogen. The degree of colouration indicates the amount of nitrogen present in the leaf.

Dark blue : Sufficient

Light blue : Slightly deficient

No colour : Highly deficient

2. Posphorus

Reagents: Ammonium molybdate solution.

8 gm of ammonium molybdate is dissolved in 100 ml of distilled water. To this, add 126 ml of conc. hydrochloric acid (HCL) and volume is made upto 400 ml with distilled water. This stock solution is kept in an amber coloured bottle and at the time of use, it is taken and diluted in the ratio of 1:4 using distilled water.

A tea spoonful of freshly chopped leaf bits are taken in a test tube and 10 ml of diluted reagent is added and shaken for 30 minutes; After shaking, a pinch of stannous chloride is added. Colour development is observed.

Dark blue : Sufficient P

Bluish green : Slightly deficient P

No colour : Highly deficient P

3. Potassium

Reagent: Sodium cobalt nitrate reagent

Take 5 gm of cobalt nitrate and mix with 30 gm of sodium nitrate in 80 ml of distilled water. To this, 5 ml of glacial acetic acid is added and the volume is made upto 100 ml with distilled water. Dilute reagent prepared (5 ml) with 15 mg sodium nitrate to 100 ml using distilled water.

Finely cut leaf bits are taken in a test tube and 10 ml of diluted reagent is added and shaken vigorously for a minute. A change in turbidity is observed.

Turbidity – Sufficient K   Less turbidity – Slightly deficient K.

4. Iron

Reagents: Conc. H_2SO_4, Conc. HNO_3, ammonium thiocyanate and amyl alcohol.

0.5 gm of the material to be tested is taken in test tube and 1 ml of conc. H_2SO_4 is added and allowed to stand for 15 minutes. After that 10 ml of distilled water and 2-3 drops of Con. HNO_3 are added. After 2 min, 10 ml of this solution is taken and add 5 ml of 20% ammonium thiocyanate solution in amyl alcohol.

Brick red : Sufficient Fe

Faint colour : Deficient Fe

5. Chloride

Reagent: 2 ml of N/50 Silver nitrate solution and add 3 drops of conc. HNO3.

Leaf bits are taken in a test tube and add 10 ml of the reagent mixture. Development of turbidity is observed.

Turbid – Sufficient chloride

Colourless – No chloride/deficient status.

CHAPTER 4

Deficiency Symptoms

4.1. Deficiency symptoms in different plant parts

Nutrients are important for growth and productivity of crops. Apart from the N, P, K, Ca, Mg and S, micronutrients such as Zn, Fe, Cu, B, Mo, Mn, Cl, etc. are also necessary for maintaining normal health and metabolism of plants. However, nutrient deficiency can lead to poor and stunted growth of plants resulting in poor yield performances. The depletion of nutrients from the soil is increasing due to adoption of modern agricultural technology through intensive and extensive cultivation of high yielding varieties. Nutrient deficiencies are very common in horticultural crops rather than in field crops.

Identification of nutrient deficiencies in crop plants through visual symptom is the cheapest diagnostic technique. When a nutrient element insufficiency occurs, below the critical level, visual symptoms may or may not appear, although normal plant development will be slowed. When visual symptoms occur, such symptoms can frequently be used to identify the source of the insufficiency.

4.1.1 Symptoms

Crops manifest through distress signals, if there is any deficit of water or invasion of an organisms namely pest and diseases or shortage of nutrients. Shortage or deficiency of a particular nutrient element disturb the metabolism and thus produces the characteristic deficiency symptoms depending upon the deficient element. Symptoms of nutrient deficiencies have been used for many years to diagnose the deficient elements. These symptoms can be classified in to five types.

1. Chlorosis, which is a yellowing of plant tissue (either uniform or interveinal), due to a reduction in the chlorophyll formation process.
2. Scorching, wilting, necrosis, or death of plant tissue
3. Lack of new growth or terminal growth resulting in rosetting.
4. An accumulation of anthocyanin and an appearance of a reddish colour and
5. Stunting or reduced growth with either normal or dark green colour or yellowing.

Deficiency as such can be classified in to two types depending upon the cause of the deficiency.

(1)Absolute deficiency

Deficiency due to the absence of a particular element or low availability in the soil is termed as absolute deficiency.

(2) Induced deficiency

Nutrient deficiency which occurs in crops growing in soils with adequate nutrient content due to other factors which limit the availability or the uptake or translocation of nutrients is called as induced deficiency.

1. The first step in identifying nutritional disorder by visual symptom is observation on the growth and development of the plant. Growth can be stunted by the deficiency of all elements.
2. The second step is to note, which plant part is affected, whether the foliar symptom appears on the lower leaves or older leaves on the stem or flowers.
3. After identifying the part of the plant, the third step is recognition of the nature of the symptoms – whether chlorotic, necrotic or deformed.

The deficiencies are manifested in various plant parts. The expression of deficiency symptoms are varied among the plant parts Viz., whole plant, leaves, terminal shoot, stem, roots, flowers, fruits, seeds, etc., depending upon the role of the element, nutrient content, mobility of ions and sensitiveness to the particular element.

4.1.2 Whole plant

A good diagnostician must be broad in his knowledge and meticulous in his approach. For deficiencies of N, K, P, Cu and Ca, one should observe the whole plant.

- Nitrogen hunger signs show general yellowing in older leaves and stunted plant growth with poor yield.
- Potassium deficiency has been characterized by bushy growth, early sprouting of axial buds and susceptible to lodging by wind or rain. Sometimes, plant may also show general wilting in older leaves.
- Phosphorus deficiency has been characterized by stunted plant growth and delayed maturity (late flowering)
- Calcium deficiency shows general wilting in younger parts of the plant.
- Copper deficiency leads to production of numerous water shoots in fruit trees.

4.1.3 Leaves

Plants speak through their leaves. They express any distress signals through leaves. Similar to human beings, where the face is the expression of the mind, in plants too, healthy leaves are the expression of good health of a plant. The deficiencies of N, K, P, Mg, S, Zn, Cu, Fe, Mn, B, and Mo are well manifested in leaves.

- Uniform chlorosis of whole leaves is due to the shortage of N. Reddening of older leaves in Eucalyptus is also due to the deficiency of N.
- Phosphorus deficiency leads to unusually thick leaves. Leaves become narrow with purple colouration on the lower side of older leaves.
- Potassium deficiency has been characterized by leaf shedding, dropping of leaves, leaves thickening with pointed necrosis, however, veins are not affected. First the older leaves dry up followed by younger leaves. Development of small necrotic spots on the leaves commencing from margin to midrib and further to tip downwards. K deficiency is considered to be one of the factors responsible for reddening in guava.
- Inverted 'V' shaped chlorotic pattern of older leaves is due to the deficiency of Mg. The lamina near the petiole and the margins near the base remain green. The other parts of the leaf starting from the tip become chlorotic, so also the margins. The leaves become red in colour in *Hibiscus* sp.
- Sulphur deficiency leads to general chlorosis of younger leaves and in acute cases it spreads to the other leaves and the deficiency symptoms are similar to N deficiency.
- Zinc deficiency results in rosetting, interveinal chlorosis of young leaves, puckering of leaves, eruptions in leaf surface and leaves become small.

- Iron deficiency is expressed as interveinal chlorosis of younger leaves. The leaves remain green and the lamina in between the veins become yellow. In severe cases, the veins also become yellow and the whole leaf has a bleached appearance.
- Copper deficiency shows unusually thick leaves, brittle in nature, chlorotic patches, cupping of leaves, mishappenly twisted leaves. One should not look for cupping of leaves alone for Cu deficiency, because drought, nematode attack also causes cupping of leaves.
- Manganese deficiency results in interveinal chlorosis of younger leaves. The midrib, secondary and tertiary veins remain green and the remaining portions become yellow and the leaves have a netted like appearance. Mn deficiency is also one of the factors responsible for reddening of guava.
- Boron deficiency results in unfurling of terminal leaves and puckering of leaves. The reddening or guava is also due to the deficiency of B.
- Molybdenum deficiency is more prominent in cole crops than the other crops. "Whiptail" in cauliflower is the typical symptom of Mo. The leaf will have midrib alone with reduced lamina adjoining the midrib. The leaves appear as malformed, asymmetric with soft chlorosis.

4.1.4 Terminal shoots

Terminal shoots are the place of peak physiological activity in any plant. Deficiency or insufficiency of nutrient will immediately be reflected in the terminal shoots. The deficiency of K, Ca, and B are clearly expressed in the terminal shoots.

- Potassium deficiency leads to weak stalks.
- Calcium deficiency results in inhibition of terminal bud development.
- Boron deficiency has been characterized by death of terminal shoots and stem and in acute cases the death of the entire growing point with necrosis of internal tissue.
- Rosetting – reduction in internodal length of terminal shoots is due to the deficiency of zinc.

4.1.5 Stem

Stem or trunk exhibit the symptoms of some of the elements. One should look carefully and closely for deficiency symptoms on the bark of fruit trees.

- Phosphorus deficiency induces stunted plants but have stronger stems.
- Sulphur deficiency produces thin and erect stem.
- Potassium shortage leads to stocky appearance with short internodes.
- Magnesium hunger signs results in sharply delineated, sunken dark to blackish- brown necrotic blotches on the stalks in grape vine.
- Copper deficiency results in slender and weak stems with poor lignification, splitting or cracking on the barks.
- Boron shortage produces cracking on the bark with gum exudation.

4.1.6 Roots

Some of the deficiencies of nutrients are also expressed in roots.

- Roots are long with poor ramifications in N deficient plants.
- Phosphorus deficiency shows poorly developed fibrous roots. Sometimes roots become swollen in nature.
- Potassium deficiency induces susceptibility to root rot diseases.
- Death of root tips is due to Calcium deficiency.
- Root tip stops growing and side root proliferation are due to the deficiency of Fe.
- Boron deficiency has been very well expressed by roots; death of root apical meristem; root tips swollen and discoloured; gummy exudation from the roots.
- Reduced root lengths, thickened and swollen near the tips are due to Cl deficiency.

4.1.7 Fruits

Flowering, fruit set, fruit development and ripening are affected similar to other plant parts due to the deficiency of nutrient element.

- High flower drop is due to the deficiency of Zn.
- Poor fruit set is caused by hunger signs of B.
- Fruits of uneven size with undulating surface, gummy exudation, less taste of fruits are symptoms of B deficiency.
- Cracking of fruits has been caused by the deficiency, of B and K. The concentric cracking is due to B; whereas radial cracking is due to K.
- Manganese and Iron deficiency produces poorly coloured fruits.

4.1.8 Seeds

Seeds are the ultimate end product of any plant. The seeds produced by the nutrient deficient crops are lacking one or more attributes of good quality seeds.

- Potassium deficiency results in reduced endosperm content in the seed (small seeds).
- Phosphorus and Calcium deficiency reduces the viability of seeds.
- Boron deficiency induces ill filled seeds with poor viability.

4.1.9 Indicator fruit crops

The genetically determined nutrient uptake, translocation and utilization of the different species and varieties play an important role in diagnoses of nutrient deficiencies. This property results in identification of indicator plants of particular deficiencies. These Indicator plants help to diagnose the nutrient deficiency very easily. These plants are more sensitive or susceptible to particular deficiencies and the deficiency symptoms manifest themselves well ahead of other crop species. Indicator plants produces deficiency symptoms very clearly and characteristic to that particular deficient nutrient elements.

The fruit crops that are more demanding with regard to a particular element, and its insufficiency causing the most characteristics deficiency symptoms are known as indicator plants. In these plants, the typical symptoms shown are unlike those for other deficiencies on the plants, and these plants are markedly susceptible to the particular deficiencies. A list of fruit crops which are susceptible to particular nutrient is given below (Table 3).

Table 3: Indicator fruit crops

Nutrient	Indicator Fruit crops
N	Apple, plum
P	Gooseberry, apple
K	Gooseberry, red currant apple
Ca	Apple
Mg	Gooseberry, black currant apple
Fe	Apple, raspberry, pear, plum
B	Apple, grape
Mn	Apple, cherry, raspberry
Cu	Pear
Zn	Apple, pear, citrus
Mo	Citrus

4.2 Defeicency symptoms of nutrient element

4.2.1 Nitrogen

Nitrogen is the most important nutrient element in crop production. It is an element circulating between the atmosphere, the soil and living organisms. In 1804, the essentiality of N for plant growth was discovered by De saussure. Fruit crops require large amounts of nitrogen. Besides the N taken up by the plants, much of this element is also lost through leaching, erosion or escape as ammonia or elemental N into the atmosphere. Such losses cause great demand for the Nitrogenous fertilizers. Usually the Nitrate content of the soil solution is of the major importance in crop production. Nitrate levels in the soil solution can be as high as 20 to 30 mM after nitrogen fertilizer application and nitrate ions are rapidly taken up by roots depending upon the demand.

4.2.1.1 Nitrogen in soils

Unlike other elements, N is not present in parent soil minerals. Hence, it has to be added to soil from other sources. One of these sources is the organic matter, originating from both the plants and animals. Nitrogen constitute about 5-6 per cent of soil organic matter by weight. Nitrogen is added to soil by free living and symbiotic bacteria. Rainfall is also responsible for the addition of some N to the soil. The top layer of cultivated soils usually contains 0.02 to 0.4 per cent N, the quantity depending upon the soil type, temperature and rainfall. Indian soils are usually deficient in N, our climate, cropping pattern and soil management practices further aggravate the situation.

The categories of available N status of the soil is given below.

Rating	Available N (Kg/ha)
Low	<280
Medium	280 – 450
High	>450

4.2.1.2 Factors affecting N availability

1. The activity of both free living and symbiotic bacteria influence the N content of the soil through biological N_2 fixation.
2. Soil pH influences the activity of nitrifying bacteria.
3. Adequate supply of O_2 ensures the activity of nitro bacteria.
4. Nitrogen mineralisation is reduced at higher soil water content.
5. Nitrification is also influenced by temperature. Temperature at about 30°-35° C is found to be optimum.
6. Sandy soils with low CEC permits appreciable movement of ammonia nitrogen in to the subsoil.

4.2.1.3 Physiological role of Nitrogen

Nitrogen is the fourth most abundant element in plant after C, H, and O. Nitrogen usually accounts for 1-5% of whole plant dry weight. Nitrogen is an indispensable elementary constituent of numerous organic compounds viz., amino acids, nucleic acids, proteins, nucleotides, chlorophyll and numerous secondary substances such as alkaloids. Nitrogen is also an important constituent of protoplasm and those cell constituents responsible for the storage and transfer of genetic informations like the chromosomes, genes and ribosome's. Nitrogen also acts as a constituent of enzymes, it is implicated in all the enzyme reactions taking place in the cells and thus plays an active role in energy metabolism. Nitrogen is absorbed by plants in ionic form of NH_4^+ and NO_3^- through roots or leaves. Nitrate is often a preferential source of crop growth but much depends on plant species and the environmental factors.

The uptake of NH_4–N takes place best is a neutral medium and it is depressed as the pH falls. The converse is true for NO_3 absorption, a more rapid uptake is occurring at low pH values. Gaseous ammonia may also be absorbed by the upper plant parts via the stomata. Nitrate-N should be reduced to ammoniacal form before it is assimilated. Reduction process occurs in chloroplast through the action of nitrate reductase. In green plants, protein N is by far the largest N fraction and amounts to about 80-85% of the total N. The N of the nucleic acids makes up about 10% and the soluble amino N about 5% of the total N present in plant material. Many crops are cultivated essentially to produce plant proteins. In vegetative plant material, the protein are mainly enzyme proteins; where as in seeds and grains, the major protein fraction is made up of storage proteins. Nitrogen is also an essential constituent of various enzymes. As much as 70 per cent of total leaf nitrogen may be in chloroplast, suggesting that in a leafy plant roughly half of the total nitrogen may be in plastids.

Nitrogen taken up by plant roots is translocated in the xylem to the upper plant parts. Nitrate and amino acids are thus the main forms in which N is translocated through xylem sap into the vascular system of higher plants. About 70-80% of amino acids present are rich in N, with N/C ratio greater than 0.4. It is believed that the function of these N–rich molecules (glutamine, asparagine) is to transport N with a minimum amount of carbon.

Being a part of proteins, it is an important constituent of protoplasm. Enzymes, the biological catalytic agents which speed up life processes, have N as their major constituent. Nitrogen is also influencing the phytohormone balance. The synthesis of plant hormones viz., cytokinin and gibberellic acids are encouraged by nitrogen.

4.2.1.4 Nitrogen requirement of fruit crops

Generally inorganic fertilizers are applied one year after planting. The quantity of nitrogen fertilizers to be applied to young fruit tree may vary from crop to crop. Recommended level of N for different fruit crops (bearing trees) are given bellow.

Table 3: Nitrogen requirement of fruit crops

Crops	Quantity of N for bearing plant(g/plant/year)
Mango	700
Acid lime	600
Sweet orange	800
Grapes (mascat and seedless)	400
Sapota	400
Guava	500
Pomegranate	625
Date palm	600
Apple	500
Peach	600
Pear	400
Strawberry	150 kg/ha
Cherries	225
Oil palm	500
Cashew nut	500
Coconut	560
Walnut	60kg/ha

The above recommended level of N should be applied in 2 or 3 splits depending upon the crops viz. two splits for mango viz., June- July and October, for banana 3rd , 5th , and 7th month after planting. Depending upon the splits, $^1/_2$ or $^1/_3$ of N, full dose of P, full dose of K and FYM @ 25 – 50 kg/ tree should be mixed and applied at 30 – 100 cm away from the trunk / plant and incorporated in the soil. Generally, N is applied just prior to the flower initiation as first split and second split at the time of fruit growth. Hence, the time of N application in spilts depend upon the crop phenology and the distribution of rainfall. To overcome the N deficiency of the standing crop, foliar application of urea is recommended. Foliar application of urea is of much more helpful to save the crop from acute N deficiency. Urea @ 1-2% concentration is applied on the foliage of the tree. One important precaution is that the biuret content of the urea should not exceed 2.0%. Otherwise there will be leaf scorching.

4.2.1.5 Diagnosis of Nitrogen deficiency

The Indian soils are being low in N, the fruit trees grown under such conditions usually show signs of N deficiency. Owing to the great importance of N for plant growth, development and yield, deficiencies of this element are rapidly and unmistakably shown by reduced growth rates and the decomposition of chlorophyll. These symptoms are now widely known and can be easily recognized. The inhibition of chlorophyll synthesis in plants due to lacking of N leads to pale to pale yellowish green appearance and the deficiency becomes more be severe, the colour gradually turns into yellow, which is otherwise called general chlorosis. Ultimately, the leaves start to wither, starting at the leaf tips and margins, and dry out, turning yellowish brown to brown. Therefore, the signs of N deficiency first appeared as yellow discolouration and withering of older leaves and parts of the plant, the younger parts remaining green for some time. Plant lack N remains more or less stunted, depending upon the severity of the deficiency. Longitudinal and shoot growths are inhibited, as also is growth in thickness, the erect appearance of the stems and leaves (upright leaves close to the stem) and shortened leaves. This differs from the P deficiency in which the leaf tip droops slightly. Owing to the relative surplus of carbohydrates in deficient plants, which is common in some plant species, it comes to orange red, red to purplish-red discolourations. Sometimes N deficiency can be induced temporarily. Prolonged drought, inundation owing to heavy rainfall or excessive irrigation can lead to temporary appearance of nitrogen deficiency symptoms even on soils that have received liberal applications of nitrogenous fertilizers. The following examples can be regarded as typical N visual deficiency systems of fruit crops.

Table 4: Critical level of N content of some important fruit crops

Crops	Concentration (%)
Apple	*1.7*
Pear	*1.8*
Citrus	*2.5*
Grapes	*2.0*
Guava	1.6
Papaya	*1.4*
Pineapple	1.4
Pomegranate	*1.1*
Peach	*2.5*
Plum	*2.1*
Banana	*2.5*
Walnut	*2.1*
Litchi	*1.3*
Mango	*1.8*
Avocado	*1.8*
Oil Palm	*2.5*
Coconut	*2.0*
Cashew nut	*1.4*
Strawberry	*2.0*

Plant analysis indicates the concentration of nutrient element in plant tissue. Plant analysis can aid in diagnosis. Selecting the right plant part, stage of growth and the time of sampling are very crucial in plant analysis. The critical level of nitrogen varies from crop to crop. The critical content of nitrogen can be used for understanding the deficiency which can be used to confirm the visual deficiency systems of nitrogen.

Banana

Nitrogen deficiency results in stunting of plant growth and reduction in the rate of leaf production. New leaves become progressively smaller and pale in colour. The whole plant is pale green yellow colour. Leaf stalk become green yellow pink with a translucent appearance and mottled with brown, while the margins were pronounced with purple colour. Leaf sheath become pinkish yellow with dominant pink areas. It is also reported that the initial stages of N deficiency are characterized by pinking of undersides of midribs and leaf petioles due to rapid development of anthocyanin pigments. Plant growth is stunted and a shortening of distance between leaves results in a "rosette" appearance with the older leaves becoming chlorotic.

Grapes

When N becomes limiting, the small, thin, stiff leaf blades turn pale green and then yellow. Young shoots, petioles and cluster stems become pink or red regardless of the cultivar. Under drought conditions, the leaf margins rolls slightly upward, wilt and dry up. Sometimes, light brown island of dead tissue appear between the main veins of the basal leaves and in extreme case, the pale leaf blades may wither and abscise. Due to N deficiency, berries start to mature early and do not attain their normal size. In commercial orchards, the first symptom may appear after beginning of ripening because N is being translocated from the leaves to the berries.

Pine apple

Due to N deficiency, plant growth is observed to be very poor with no production of suckers, slips and fruits. Root growth is limited and only a few roots are formed. Leaves become yellow with red being more predominant at the midrib. Nitrogen deficiency also leads to delayed fruiting and deformed fruits. Older leaves become pale yellow to orange; whereas emerging leaves were small, pale green and had developed necrotic margins.

Papaya

Nitrogen deficiency results in stunted plant growth and general yellowing of leaves

Citrus

Nitrogen deficiency has many manifestations. Under continuous stress for N, the trees maintain substantial growth by recycling N from older to newer leaves. Thus, the life of a leaf may be shortened from 1-3 years to 6 months or so. As the leaf matures, leaf lose N to younger leaves, turns yellow and abscises. Twigs may die back and fruit production is limited to small sixes of inferior quality. Tree growth is uneven and the tree have a dense appearance. The small pale fruits ripen prematurely and have a thick roughly textured peel.

Guava

The plant growth is retarded, leaves become uniformly chlorotic, remain small in size and relatively more upright. The purple patches on the leaf are concentrated more on both sides of the midrib and principle veins. As the deficiency advances, the colour of patches turn to brown and the tissues broke away leaving holes in the lamina.

Pomegranate

Crop growth is checked and only few of their branches develop; leaves become uniformly chlorotic with light green midrib and veins and in extreme cases necrotic spots develop on the chlorotic margins.

Sapota

General yellowing and reduced vegetative growth. Yellowing of leaves commences from leaf margin, which spreads towards midrib and finally leaves become chlorotic.

Datepalm

General yellowing of older leaves and retarded emergence of new leaves.

Coconut

Yellowing of older leaves, resulting in an unhealthy appearance of the crown. In advanced stages, young fronds also turn pale green giving the leaflets a dull appearance. The severely discoloured old leaves may become golden yellow. Many bunches abort and there is a reduction in the number of female flowers on the inflorescence. The palm shows stunted growth and the stem tapers upwards to give a "pencil point" appearance. Reduction in size and number of leaves also noticed.

Oil palm

Nitrogen deficiency is expressed in uniformly pale, yellow green leaflets with a sharply reduced growth rate. Midrib tissues become bright yellow and the symptoms are equally pronounced on both upper and lower rank pinnae.

Goose bearies

Nitrogen deficiency leads to restricted shoot growth in the first season followed by a reduction in the size of berries and the total crop in the second season. The blossom development is restricted and followed by early defoliation. Root growth was also effected. The leaves become unfolded slowly and are pale green. Later, red and yellow tints could be seen all over the leaf blade.

Litchi

Leaflets are small and chlorotic with slightly curling margins. Defoliation occurrs in both the older leaves and latest flushes of growth. Often the newest leaves failed to mature and the new terminals were leafless.

Mango

The plant growth is stunted. The fruit set is greatly reduced and leaves mature without attaining normal size. The leaf stalk tend to form narrow angles with the stem. The mature leaves become yellow. The first symptom appear near the base of the current growth and progress towards the tip. Twigs become stiff and woody, limited in length and diameter. Under acute deficiency, the fruit size is greatly reduced, tend to ripe and drop prematurely.

Cashew

Deficiency symptoms of nitrogen always appear first in older leaves, stem becomes thin and growth is stunted. The leaves are so smaller than those of other plants. Nitrogen deficiency starts appearing 7 weeks after planting. Colour of the leaves change gradually from dark to pale green and to yellow, coincided with stunted growth and ultimately the seedling will die after 4 months in the case of severe N deficiency.

Apple

Leaves are small, narrow and pale green, older leaves having a yellowish orange, reddish or purplish colour and falling prematurely. The stalks form an acute angle relative to the twigs, which are short and thin. Shoot growth is inhibited. Few buds are formed, the blossom is sparse, the period of fertilizability is shortened and the fruit colour is not well developed.

Pear

General yellowing of leaves, reduction in vegetative growth and low fruit production are observed. Shoots are thin and weak, fruits have pale colour and become woody in acute deficiency.

Avocado

Nitrogen deficiency results in stunted plant growth and the bark is reddish brown and the spring flush shoots are short and thin. On elongation of shoots, the immature leaves become amber to abnormally bright red in colour. The mature leaves are very small, stiff and yellowish green. A progressive burning from tip to the base of the leaf develops in the summer and fall on spring, flush leaves and the leaves abscise when nearly all the blade had burned.

Apricot

The leaves are short, yellowish-green and the twigs are thin. The shoots often blossom profusely, but the few fruits that the trees bear will be small.

Pecan

Leaves lose green color, changing to yellowish-green; reddish-brown necrotic spots appear; shoot growth retarded and shoots finally show die back symptoms.

Strawberry

Strawberry plants show stunted growth and have yellowish-green leaves. The leaves are small and have rigid upright stalk. The plants produce few runners. Older leaves turn bright orange-red to red starting at the margins. The teeth of the leaf margin often turn brown and die before the whole leaf is discoloured. The veins at the leaf bases retain their green colour for sometime. Flowering and frutification are more or less reduced, and the fruits are small.

Aonla

Plants starving of nitrogen put up poor vegetative growth and remain pale and unproductive.

Peach

Trees stop growth and the leaves become pale green. Reddish tints gradually appear on leaves and branches are reddish-brown with spots, some laminae are small and often twisted and curved badly. Premature defoliation occurred beginning with older leaves. Leaves turn to an orange or reddish-yellow colour prior to abscission. Many fibrous roots are produced. The root-top ratio of deficient trees is three times greater than the healthy trees.

Raspberry

Initially, the symptoms appear as uniform yellowing of older leaves, later turning to bright yellow and necrotic at tips and margins before falling. As the deficiency progresses, younger leaves are also affected. The growth of cane is restricted and they become stiff. Red colouration may develop both on the canes and the leaves.

Plum

The deficiencies of N cause decrease in vigour and reduction in yield. The colour of the leaves become pale yellow and fruits mature early.

Walnut

Leaves may be pale in colour and small sized with reduced shoot growth. Leaves senescence early, become yellow and defoliate earlier than normal. Sometimes, N deficiency also results in lack of vigour, suppression of shoot growth and dying of small twigs, reduction in fruitset and small poorly filled nuts.

Blueberry

The deficiencies of N cause early cessation of shoot growth and uniform paleness of the foliage, a characteristic symptom of N deficiency commonly observed in the field.

4.2.1.6 Correction of N deficiency

Nitrogen deficiency in fruit crops can be rectified by timely application of required quantities of nitrogenous fertilizers. Generally the amount, the source of N, the time and the method of application depends upon the climate, soil and the type of fruit crops.

4.2.1.7 Effective N management

1. To reduce leaching of N, nitrate fertilizers must be avoided in coarse textured soils and soils under high rainfall. Split application of N fertilizers as well as application of slow release N fertilizers may be encouraged. The leaching loss of N is estimated to be around 70% of the added N.
2. To reduce denitrification of N, nitrate fertilizers should be avoided and application of ammonium sulphate and urea should be encouraged. Split application as well as application of slow release N should be adopted. The denitrification loss of N is estimated to be around 87% of added N.

3. Continuous application of high rates of ammonium sulphate causes loss of calcium in surface soils (ploughed layer) and increases the acidity. Hence, at least once in three years, the available Ca content and the pH of soil should be determined and lime should be applied.

4. Decomposition of nitrite occurs very fast in strongly acid soils (pH<5) with high organic matter. In order to reduce decomposition loss of N, the soil should be well aerated by tillage and acid soils should be limed to pH 7. Nitrate fertilizers can be recommended. Foliar spray of urea may also be followed.

5. Weeds compete for N with crop plants. Hence, to increase N efficiency of crop plants, weeding should be done at the early stages of crop growth.

6. Application of high doses of urea and ammonium sulphate in bands and application of particles of urea of large size increases ammonia concentration in soils that raises soil pH to above 7.0 and affects the microbial activity particularly, nitrobacter activity. Hence, nitrite accumulation (NO_2^-) in soil becomes toxic to seedlings. In order reduce the toxic effects, urea, ammonium sulphate or diammonium phosphate should be placed at least 2.5-3 cm below the seed and 5-7 cm away from the seed. Acidifying fertilizers should be mixed with urea.

7. Ammonia volatilization causes heavy economic loss to the farmers. The volatilization depends upon the type of fertilizers and type of soils. As regards fertilizers, the ammonia volatilization capacity of the fertilizers are graded as urea > ammonium sulphate> ammoniumchloride> ammoniumnitrate. The ammonia volatilizing capacities of the soils are graded as calcareous soils >alkaline soils > neutral soil > acid soils. In neutral and acid soils, ammonia volatilization occurs from urea only, not from ammonium fertilizers. Surface application of fertilizer increases ammonia volatilization. Incorporation of fertilizers with soil decreases volatilization. Ammonia volatilization is found to be heavy from bare or coarse–textured (sandy soils), calcareous and alkaline soils.

Ammonia volatilization from the urea added to the surface of warm and moist alkaline or calcareous soils can be completed within a week. Nearly 50% of the total volatilization occurs within first day (24 hr) of application of urea. The total loss is estimated to be 60% of added N. In acid soils , volatilization occurs to an extent of 10 –15% of the added N.

In order to minimize volatilization, urea must be incorporated in the soil immediately after application. In should not be left out on the soil surface. In coarse textured soils (sandy or sandy loam soils), urea may be applied on the surface followed by irrigations, acidifying fertilizers should be mixed with urea.

In acid soils, ammonium sulphate may be applied on the surface, but urea must be incorporated with soil. In the case of highly alkaline or highly calcareous soil, urea should be deeply incorporated in the soil. Slow release N fertilizers and split application are also encouraged besides foliar application if situation warrents.

4.2.2 Phosphorus

Phosphorus is one of the most essential nutrient elements for crop growth. Its functions can not be performed by any other nutrient and an adequate supply of P is required for optimum crop growth and production. By the 1860s, Ville had established the essentiality of phosphorus for plant growth. It's requirement for plant growth is low when compared to N and K and in quantities equal to that of sulphur. P content in crops ranges between 0.1 and 0.5 per cent. Though its requirement is low, P is absolutly essential for physiological functions of plant. Phosphorus fixation in soil is one of the common problems in phosphorus nutrition. As some of the labile phosphate is reduced immobile, phosphorus application rates should be about 10 –50 % higher than the quantity of P taken up by the crop. The rates generally applied to arable crops ranges from 20 to 80 kg/ha which depends upon the nature of the soil and the type of the crops. Crops with high growth rates, producing large quantities of organic material have a higher demand viz., maize and potato. Phosphorous is very much essential for leguminous crops to increase the N_2 fixation. Phosphorus is also essential for fruit crops. It enhances the fruit setting, increases the size of the fruit and quality besides its role in enhancing fruit maturity. Hence, understanding the role of P in fruit crops, its deficiency and correction are considered important.

4.2.2.1 Phosphorus in soils

Phosphorus in soils occur in the form of orthophosphate. The total content is in the range of 0.02 to 0.15% P. Soil P may be divided into two classes viz., organic and inorganic. The organic fraction is found in humus and organic material. It becomes involved in chemical reaction only after mineralisation by soil micro organism. Inorganic P occurs in soil in many forms. Only a fraction of inorganic P in soil solution and is in epuilibrium with absorbed P. The important P fractions of the soils are (a) phosphate in soil solution (b) phosphate in labile pool and (c) phosphate in nonlabile fractions. Phosphate in labile pool is in rapid equilibrium with phosphate in soil solution. Phosphate in the non-labile fraction is slowly released into labile pool. The labile pool of phosphorus is made up of phosphorus adsorbed on clay minerals.

The concentration of phosphorus in soil solution is very low and ranges between 0.3 and 3 ppm. The most important phosphorus containing ions is in soil solution are H_2PO_4- and HPO_4^{2-}. The ratio of these two ions in soil solution is pH dependent. At pH 5.0, HPO_4^{2-} is almost absent where as at pH 7 both phosphate species are present in equal proportions. The quantity of P present in soil solution, even in soils with a fairly high level of available phosphate is only in the range of 0.3 to 3 kg/ha. As rapidly growing crops absorb phosphate quantities of about 1 kg P/ha/day, the labile phosphate is more or less identical with exchangeable phosphate. The quantity of this fraction present in the top soil layer (20cm) is in the range of 150 to 500 Kg p/ha. For this reason, such soils are better able to buffer the phosphate concentration of the soil solution during the growing season. Hence, P concentration of the soil solution and the phosphate buffer capacity are the most important factors controlling the phosphate supply to the plant roots. The optimum P concentrations of the soil solution may be thus be low if the phosphate buffer capacity is high and vice versa. So soils which are prone to strong phosphate fixation often require extremely high phosphate fertilizers application in order to alleviate the effects of fixation. In these strongly phosphate-fixing soils, pH correction is also recommended, since phosphate adsorption is especially high at low pH level. If the quantity of available phosphate is in a normal range, the rate of phosphate application required should correspond to the amount of P removal by the crop.

The available P status of soil can be determined by olsen's method, which will be helpful to decide the P fertilization depending upon the level of the deficiency

Rating	**Available P (Kg/ha)**
Low	<11
Medium	11-22
High	>22

4.2.2.2 Factors affecting phosphorus availability

(A) Soil factors

1. Soil texture:. Responses to fertilizer P tend to be greater on sandy soils than those containing more silt and clay. Fixation of soil P increases as the soil clay content increases. Hence, higher P most be applied to those soils inorder to increase the P availability to plants.

2. Soil aeration and compaction: If the soils are compact, pore space is diminished, oxygen is limited and P absorption suffers.

3. Soil temperature: Low soil temperature depresses P availability and plant uptake. Lower temperature reduces the rate of mineralisation of soil organic P because of lowered microbial activity.

4. Moisture stress also reduces P availability and uptake. Low soil moisture has been found to decrease P availability.

5. Generally, higher soil organic matter levels are related to greater P availability.

6. Soil pH has an important role in P availability and affects the efficiency of applied P. Availability in most soils is at a maximum in the pH range of 6 and 7.

7. Nutrient interaction: Crop responses to P are affected by availability of other nutrients. When P availability is high, Zn uptake is reduced. Phosphorus fertilizer absorption and use efficiency by the crops are improved by the presence of ammoniacal nitrogen in the soil with P.

8 Incorporation of crop residues increases microbial activity and immobilized P is gradually released for plant use as residues decompose.

9. Roots affect biology of the soils by providing energy sources for microbes – Phosphate solubilising bacteria.

10. Mycorrhizae increases the P absortion through plant – fungus association. Mycorrhizal infection may be severely reduced by fallow periods in crop production, which increases the importance of starter P fertilization in crops grown in a fallow rotation. Prolonged wetness and flooding can decrease the mycorrhizal inoculum levels and P availability is decreased.

11. Most available P is present in the surface soil and helps concentrate roots in the zone. Phosphourus availability in soil is less under moisture stress condition. Hence, higher the root length and density, maximum will be the P absorption.

12. Hybrids and high yielding varieties require more P.

13. Since P is an immobile element in soils, placement has tremendous effects on crop responses to applied P.

14. P availability also depends upon the selection of phosphotic fertilizers which depends upon the soil type, pH and the nature of the crop.

15. Low temperature is also considered as an important factor contributing to phosphorus stress in young plants, since cool soils inhibit root growth and reduced the rate of diffusion of phosphorus to the roots.

4.2.2.3 Phosphorus in plants

Phosphorus generally accounts for about 0.1 –0.5% of the dry weight of plants. Phosphorus enters the plant through root hairs, root tips and the outermost layers of root cells. Phosphorus is taken up mostly as the primary orthophosphate ion ($H_2PO_4^-$) but sometimes is also absorbed as secondary orthophosphate ion (HPO_4^{2-}), this later form increasing as the soil pH increases. Once enters the plant root, P may be stored in the root or transported to the upper portion of the plant. It has been reported that 80% of the absorbed P to be converted to organic compound within 10 minutes after absorption. The predominant organic compounds are hexose phosphate and uridine di phosphate. Phosphate shows ready mobility in both upward and downward direction.

Plant roots are capable of absorbing phosphate from solution of very low phosphate concentrations. Generally, the phosphate content of root cells and xylem sap is about 100 to 1000 fold higher than that of soil solution. This shows that phosphate is taken up by plant cells against a very steep concentration gradient. The rate of P uptake also depends upon the pH. At pH 4.0, P uptake is 10 fold higher than that of 8.7. The rate of phosphate uptake declines rapidly with increasing pH. The downward translocation of P is mainly through phloem in the form of phosphoryl coline. Young leaves acquire P not only from roots but also from older leaves.

4.2.2.4 Phosphorus in physiology

Phosphorus in the plant occurs in inorganic form as orthophosphate and to a minor extent as pyrophosphate. The organic forms are phosphate esterified with hydroxyl groups of sugars and alcohol's viz., phosphate ester of fructose 6 phosphate. Such compounds are intermediates in metabolism. Phosphorus is also bound to be lipophilic compounds – phospholipid viz., lecithin. Phosphorus is involved in metabolism in its formation of pyrophosphate bonds which allow energy transfer viz., CTP. UTP, GTP and ATP. All these are involved in the synthesis of nucleic acids viz., RNA and DNA. The phosphate group in nucleic acids bridge the ribose with another ribose by two ester bonds. Hence, phosphorus is a vital component of the substances that are building blocks of the genes and chromosomes. Phosphorus is also a component of coenzyme NADP, NADPH, AMP, ADP and ATP. These are considered as most important and universal transformers of energy in the cell.

An adequate supply of P is essential to the development of new cells and to the transfer genetic code from one cell to new cells. Pyrophosphates are also important for citric acid cycle and protein synthesis.

Another organic P compound is phytin. It is the calcium and magnesium salt of phytic acid which accounts for upto 80% of the total phosphorus in grain and seeds and is considered to be one form in which phosphorus is stored. Immediately after pollination, there is an increase in P transport towards the young developing seed. During the seed germinarian, the phosphate is converted into other form and used in the metabolism of young plants. This explains how phosphorus deficiency reduces germination ability of seed and the further growth of the seedlings.

Most of the phosphate present in roots, stems and leaves is in inorganic form. The proportion of inorganic P of the total P is the highest in older leaves. Younger leaves contain relatively high quantities of inorganic P predominately in the form of nucleic acids. Therefore, actively growing shoot require relatively higher amounts of phosphorus. Phosphorus is also important for reproductive development. Lack of P results in delay in flowering and ripening. More P is needed during fruit growth and seed formation. Phosphorus is considered as most essential for root growth. It is also essential for photosynthesis. Phosphorus is particularly important for leguminous plants for N_2 fixation through rhizobium.

4.2.2.5 Diagnosis of phosphorus deficiency

When P is limiting, the most striking effects are reduction in leaf expansion as well the number of leaves. Shoot growth is more affected than root growth which leads to a decrease in shoot/root ratio. Root growth is also reduced by P deficiency leading to low roots mass. Generally, the symptoms of P deficiency appear in older leaves. A dull dark to bluish – green colour is the characteristic of plants lacking phosphorus, particularly in the early stages.

Phosphorus deficiency inhibits reactions in the citric acid cycle, thus leading to the accumulation of pyruvic acid. Since protein synthesis is also inhibited, concentrations of non-proteinaceous compounds also raises. The inhibition of starch and cellulose synthesis in phosphorus deficient plants leads to abnormally higher sugar levels and these inturn encourages the synthesis of anthocyanins. Reddish violet or deep purple discolorations develops depending upon the basic colour of the leaves. Further, symptoms accompanying prolonged deficiencies include the appearance of brown to dark brown necrotic spots on older leaves, darker brown necrotic lesions at the leaf margins and the death and felling of older leaves. Plants are with small leaves and poor lateral growth. Premature defoliation of leaves, reduction in blossoming and delay in bud opening are also due to P deficiency.

Plants lacking phosphorus are small, tender, stunted but stronger stem. Owing to the increased synthesis of anthocyanin, reddish, reddish-violet or deep purple colouration develops depending upon the basic leaf colour.

Fruit trees show reduced growth rates of new shoots, and frequently the development and the opening of bud is unsatisfactory. The formation of fruit and seed is especially depressed in plants suffering from P deficiency. Thus not only low yields but also poor quality fruits and seeds are produced. The leaves of P deficient fruit trees are frequently tinged with brownish colour. Such leaves fall prematurely. The plant analysis is also helpful to identify deficiency symptoms. The critical content of P varies from crop to crop. The critical concentration of P for some of the fruit crops are furnished below.

Table 5: The critical concentration of P in fruit crops

Crops	Phosphorus content (%)
Avocado	0.05
Coconut	0.12
Pomegranate	0.19
Papaya	0.30
Litchi	0.14
Pear	0.11
Apple	0.17
Crapes	0.30
Guava	0.18
Citrus	0.11
Banana	0.17
Mango	0.10
Oil palm	0.15
Cashewnut	0.06
Pine apple	0.16
Walnut	0.10
Avocado	0.13
Strawberry	0.20
Peach	0.18

4.2.2.6 Phosphorus deficiency in fruit crops.

Coconut

Phosphorus shortage in plants result in poor root development and characteristic discoloration of leaves.

Oil palm

Phosphorus deficient plants may be stunted with short fronds and the palm trunk may have a pronounced pyramid shape.

Cherry

Phosphorus deficiency symptoms appear first as limited and slender growth. Young expanding leaves along the margins and main veins frequently show purplish discolouration. Lateral buds may remain dormant or die and as a result, few lateral shoot appear. Blossoming and fruiting are reduced and bud opening in the spring may be delayed. Deficient fruits ripen early, have a greenish ground colour with poor eating quality.

Apple

The foliage and stems have a dark greyish - green colour. The stalk may be bronzed or show purplish colour. Shoot growth is restricted. Size of leaves is small and they may crop early. Buds may die without any apparent reason. The fruit is dull and unattractive. It also lack firmness. Fruit yields are low.

Gooseberry

Leaves are dark green in colour, poor in growth, small in shape and size. The leaves are smaller turning to dirty greyish green in the later stages. Older leaves eventually dried to brownish colour and shed prematurely.

Raspberry

Foliage of a P deficient raspberry plant first appears unusually darker green with a shining wax surface. Following these initial symptoms, there is a reduction in leaf size, stiffening and curling of the leaves and intensification of the darker green colour which changes gradually to a coppery purple. The older leaves become mottled, develop necrotic areas and die. The roots become darker and finally broken.

Black currants

Leaves are dark green and have purple botches. The leaves fall early. The stems are thin, short and the blossom is poor. The berries ripen late and are sour in taste.

Strawberry

Due to phosphorus deficiency, the leaves turn bluish to dark green, starting from the small veins and leaf margins. The leaves droop and turn deep purplish bronze with reddish margins. The leaf stalks are intensive red. On the undersurface of the leaves, the midribs and veins are violet tinged. Flowering is poor, and few fruits develop.

Litchi

Leaflets become abnormally large and are dark green. The texture of the leaves is rough. There are dark brown scorched areas in the old leaves and drop.

Pine apple

Plant growth is poor, with no production of fruits, slips and suckers. The colour of the new leaves becomes dark – purple green without any signs of Chlorosis. Deficiency also resulted in reduced leaf number. Sometimes, older leaves had marginal chlorosis with large bronze and purple chlorotic areas. Phosphorus deficiency resulted in dark green, purple-ribbled leaves, and fruits of such plants were sour and watery.

Sapota

Phosphorus deficiency results in dark green foliage, reduced growth, premature shedding of leaves and fruits become hard and small. In P deficient plants, purplish flecks on lamina towards the tip is formed and then rusty pigmentation appear all over the leaves.

Date plam

Leaves develop dull purple –green colour.

Citrus

Phosphorus deficiency causes low assimilation rate, low vegetative growth, defective formation of buds, poor fruit setting, bending and discoloration of leaves which lose their deep green colour and fade to bronze. Lateral shoots are reduced and grow less vigorously. Lateral buds may remain dormant or die. Premature drop of fruits may be seen. In extreme cases, fruits become sour, sometimes necrotic areas develop on leaves. Young trees show sparse foliage and open appearance. Blossoming and fruiting are limited and delayed. Fruits become larger, coarse and mishappen.

Cashewnut

Phosphorus deficiency reflects on stunted growth with dull –green foliage. Lower leaves of cashew seedlings are withered and fallen.

Avocado

Phosphorus deficiency results in a reduced top growth, smaller and rounder leaves, premature defoliation, dieback and considerably reduced root growth. Leaves develop a dull brownish – green colour.

Papaya

Phosphorus deficiency results in stunted plant growth, reduced blossoming. Leaves showing bluish green colour with tint of purple on petiole.

Pomegranate

Lack of phosphorus results in over all expression of plant growth. The shoots and stems become thin, number of branches and leaves are reduced, leaves of middle and apical portions on tertiary and side shoot become chlorotic (yellow) and the basal older leaves remain dull green and loose lustre. The chlorosis starts at margins and tips of the leaves and the margins of young leaves roll.

Guava

Phosphorus shortage results in stunted crop growth. Leaf margin turn coppery (purpling) and finally whole leaf becomes coppery and drop of prematurely. In acute deficiency, purple patches enlarged, coalesced and formed a complete band of deep brown colour all round the leaf.

Grapes

The deficiency of phosphorus results in reduced growth, dull grey green small leaves, premature defoliation and premature fruit ripening. Sometimes, violet –red anthocyanin pigments appear on the petiole due to phosphorus deficiency. A complex dissorder known as "acid injury" was reported due phosphorus deficiency. The symptoms appear first on older leaves and the margin of the leaves turn golden yellow to light brown. The central part of the leaf usually remain normal green with chlorotic areas becoming necrotic late in the season and in the extreme cases the whole leaf blade dries up. Under severe deficiency, the leaves turn reddish in colour.

Walnut

Trees usually show very weak shoots and have sparse foliage with smaller than normal leaves. These leaves are yellow and develop necrosis in irregular areas followed by early defoliation.

Pear

Due to phosphorus deficiency, severe burning of the margins and tip of the leaf blades occur early in the growing season, a decrease in leaf size, failure of the fruit to develop properly, very short terminal growth, a scaly appearance of the bark and dying back of the new growth appear.

Mango

Symptoms of deficiency first appear on older leaves and brown necrotic spots appear in between the veins of leaf blade. As the time advanced, the entire leaf blade was covered by such spots developing against a yellowish background. Finally, the whole leaf became brownish, dried up and dropps. Premature defoliation is so severe except the younger leaves at the top of the plants, while all the leaves shows the symptoms of dropping.

Banana

Phosphorus shortage causes stunting of growth due to lower rate of leaf production. Leaves are somewhat choked at first green (even dark green) but rapidly acquire a spreading marginal chlorosis followed by confluent necrosis and premature death. The growth of the plant is reduced long before the visible deficiency symptoms appears.

Peach

The deficiency manifests on older leaves in the form of development of purple colour. The purple colour development starts in May and progresses slowly through June and with the onset of rains in July, it becomes more evident. The purple colour of the leaf changes to bronze and in extreme cases, the tree gives a fired look from a long distance. The upward curling of leaves, and breaking of leaf margins are the symptoms associated with phosphorus deficiency. There is premature shedding of leaves too.

4.2.2.7 Correction of P deficiency

Phosphorus deficiency can be corrected by the timely application of suitable phosphatic fertilizers either as basal or as foliar spray. Generally for the fruit crops, the recommended dose of P fertilizers are applied in 2-3 splits depending upon the crop along with N and K. The requirement of phosphorus varies with the type of fruit crop and nature of soil.

Usually phosphorus is applied either before or during sowing or at planting time. The type of phosphorus to be chosen for use is largely based on the soil reaction, relative costs and ease of application.

Table 6: Recommended doze of P fertilizers

Crops	Quantity of P_2O_5 for bearing fruit trees (g/plant/year)
Mango	200
Acid lime	300
Sweet Orange	200
Grapes (Muscat and seedless)	240
Sapota	160
Guava	500
Pine apple	60Kg/ha.
Pomegranate	250
Date palm	100
Apple	250
Pear	150
Peach	100
Straw berry	75
Cherries	300
Coconut	320
Oil palm	200
Cashewnut	125
Walnut	60 Kg/ha

Single super phosphate is commonly used as phosphatic fertilizer, because of its availability and low cost. Phosphorus is considered immobile in soils, i.e. it travels very slowly in the soils, and gets fixed before reaching the root zone of the tree. Hence, it is necessary to apply it in a band around the tree or to mix well with the soil. It must be applied as close to the roots as possible. The best method is to apply the phosphate in the soil before planting the trees. After application, the soil should be thoroughly ploughed and turned in so that the phosphorus gets mixed in the subsoil. Phosphorus can also be placed in the planting pit, before planting the trees.

For soils of neutral or alkaline conditions, phosphotic fertilizers having high water soluble phosphorus are preferable. On the other hand, such P fertilizers is not insisted for acid soils. Thus, the P deficiency can be removed by adding required quantitiy of right type of P bearing in mind the factors involved in the soil and the crop. Recovery of applied P is even lower than N; only 10-20% applied P is taken by the crop. However, in contrast to N, applied P is not lost from the soil but stored into its reserve.

Among the P fertilizers, majority of them are water soluble viz., super phosphate, triple super phosphate, mono ammonium phosphate and di ammonium phosphate. The non water soluble phosphate will be slowly available to the plant.

The uptake of phosphate in the soil mainly depends upon the root growth and root morphology. With respect to the application, for soils very low in available phosphorus, placed phosphorus fertilizer application is often superior to broadcast treatment. Placement of P ensures that a higher concentration of fertilizers comes in contact with a more limited soil volume. The fertilizers is thus able to saturate the soil phosphate adsorption capacity to a greater degree. The phosphate concentration of soil solution is thus higher to the placed zone. All P fertilizer treatments should take into account the relatively high P requirement of most crops in early growth stages.

4.2.2.8 For effective management of P

In general, the P uptake and utilization depends on the availability of P in soils. The soil P fixation should be minimized or reduced by adopting effective management practices.

1. Over liming of soils to increase the pH >6.5 increases P fixation and decreases P availability.
2. Frequent application of P in a crop rotation in highly P fixing soil increases the P fertilizers utilization efficiency.
3. Liberal application of organic matter reduce P fixation and increases P availability to the crops.

4. Localized application of P viz., band placement drill placement and plough sole placement reduces P fixation and increases P availability and also P is within reach of the roots for easy absorption.
5. Application of pelleted P fertilizers increases the P availability in soil by reducing P fixation.

4.2.3 Potassium

Potassium is one of the most important element necessary for growth and production of fruit crops. It is used by the fruit crops in considerable amounts and is regarded as having special importance for fruit crops. In fruit crops, it is estimated that slightly higher amount of potash is removed than N and 10 –20 times more than P. Unlike often nutrients, K does not form the components of any of the macromolecules. However, K is required in higher amounts and occurs primarily as soluble Inorganic salts and occasionally as salts of organic acids. The importance of K in fruit production is not only to increase the fruit yield but also to improve the quality of fruits.

4.2.3.1 Potassium in soils

Potassium is the most abundant element in the earth's crust. The average K content of the earth's crust is about 2.3%. In 1860 Von sacks, Knop discovered the essentiality of potassium for plant growth. The term 'potash' was, derived from the potassium salts obtained as "pot ashes" by burning wood in retorts or pots used as source of potash fertilizers. The potassium in the soil arise from the disintegration and decomposition of rocks containing K bearing minerals viz., feldspars, muscovite and biotite. They occur exclusively in the sand and silt fractions of the soil and occasionally in coarse clay. Potassium is almost found in the soil in the form of clay minerals such as montmorillonate, illite, vermiculite, chlorite and other minerals. Potassium is bound primarily in the clay fraction of the soil of particle size less than 2 mm. For this reason, soils rich in clay are also generally rich in K and clay soils may often have in excess of 4% total K. The main source of K for crops growing under natural conditions comes from the weathering of K containing minerals. Potassium in soils is available in three forms viz., (1) unavailable (2) Slowly available (3) Readily available. It is also represented as

Non Exchangeable K	→	Exchangeable K	→	Soil solution K
(90 –98%)	←	(1-3% of total K)	←	(1-5 of total K)

Among these, water soluble K (soil solution K) move freely with soil water and the most immediate source of K for plants. Potassium is always absorbed as K^+ ions. The water soluble K is constantly being replenished by the exchangeable

K. Thus, the water soluble and exchangeable forms of K are in equilibrium with each other. The more the exchangeable K^+ fraction is exhausted, the greater is the contribution of the non-exchangeable K^+ fraction to plant. Potassium release from the non-exchangeable fraction is higher under dry soil condition. Grasses are better able to utilize the non-exchangeable K^+ fraction than dicots. This is one important reason why grasses displace legumes in mixed swards under condition of low K^+ availability. The sandy soils, red loam soils or lateritic soils with Kaolinite as dominant clay mineral generally need higher K rate while heavy textured alluvial or black soils where illite is the dominant clay minerals with selective sorbtion sites for K and high K buffering power require less K.

The available K in soils can be estimated using flame photometer by N –N Ammonium accetates method. The available K can be classified in to low, medium and high depending upon the K content of the soil.

Rating	Available K (Kg/ha)
Low	<120
Medium	120 –280
High	>280

4.2.3.2 Factors affecting K availability

(A) Soil factor

1. In general, soils with higher CEC can hold a greater amount of exchangeable potassium.
2. Exchangeable K is the measure for predicting K availability.
3. Water stress slows down the movement of potassium to roots by diffusion.
4. Under high level of moisture or in compact soils, oxygen supply is lowered and the absorption of potassium and the mineral elements is lowered. Hence, poor aeration affects K availability in the soil.
5. Soil Temperature:. Temperature as an environmental factor strongy affects the uptake of K by plants. Low temperature slows down the diffusion of K^+ ions.
6. In very acid soils, toxic amounts of exchangeable aluminum and manganese create an unfavorable root environment for the uptake of potassium
7. Soils high in Ca++ or Mg++ reduces the availability of potassium

8. Minimum tillage and no tillage practices reduces the availability of K due to increased compaction, less aeration, low temperature and positional availability of potassium applied to the soil surface.

(B) Plant factors

1. CEC of root influences the uptake of K+ ions. Plants with roots of relatively low CEC are best to extract soil K.
2. Root type and density are the two of the major factors affecting K availability to crops.
3. Genotype difference exists in K absorption and utilization.
4. Higher plant population and close row spacing demands higher K than the normal spacing.

4.2.3.3 Potassium in plants

Potassium is vital to many physiological processes. Its importance stems from its activity in ionic form and its high acro and basipetal mobility. Its concentration in plants varies between 1.0 and 6.0% and are generally higher than those of all cations. Its concentration is higher in younger and actively growing parts. Potassium uptake of a plant gradually increases in the course of its development up to the flowering stage. Potassium is present in the cyotoplasm of the cells in ionic form which indicates that it might serve as a catalyst or cofactor for one or more of the enzymatic reaction of the cell. Potassium activates atleast 60 different enzymes involved in plant growth. It neutralizes the organic anions in cell sap and helps to stabilize the pH between 7 and 8. Potassium also regulates the transpiration by its role in opening and closing of stomata. Potassium activates the leaf photosynthesis by regulating the activities of enzymes involved in photosynthesis. Potassium plays an important role in the transport of metabolites in phloem, particularly with respect to transport in storage tissues. Potassium is also involved in the protein synthesis through transfer RNA. Potassium is also important as the main intracellular agent influencing osmotic pressure. Potassium encourages energy dependent nitrogen uptake and hence K is essential for N uptake and utilization. There should be proper N:K ratio for getting maximum yield viz., 1:1 for cereals, and fruit crops require relatively higher proportion of K when compared to N. Potassium ions are involved in the accumulation and utilization of ATP in photosynthesis. In general C_4 crops require more K than the C_3 crops. Potassium is known to increases drought tolerance, frost resistance, disease and insect resistance. Potassium is often considered as the quality element. It increases the quality of fruits viz., TSS, vitamin C, juice content and colour are some of the quality attributes enhanced by K application in fruit crops. Plants exhibits visible deficiency symptoms of K when the K status of plant is less than 0.7%.

4.2.3.4 Diagnosis of K deficiency:

The nature of K deficiency varies from crop to crop and even variety to variety with climatic and environmental conditions. Diagnosing the K deficiency at early stages of crop growth not only rectifies the deficiency but also increases the yield and quality. Infact, when deficiency symptoms become visible, it is too late to correct the K deficiency, even supplementary with high dose of potassic fertilizers to the crop. Hence, it is utmost importance to diagnose the K deficiency at an early stage of crop growth. Plant analyses indicate nutrient content at a particular stage. Potassium deficiency does not immediately result in visible symptoms. At first, three is only a reduction in growth rates (hidden hunger) and only later do symptom occur. The typical deficiency symptoms of K is expressed as follows. Appearance of white, yellow or orange chlorotic spots or strips on older leaves usually starting from the leaf tips and margin. In some species, irregularly distributed chlorotic spots appear, but in all cases, symptoms start from the leaf tip. The base of the leaf usually remains dark green. The chlorotic area becomes necrotic. The tissue dies and leaves dry up. Roots of K deficient plants are poorly developed and often affected by rot. Diseases incidence increase, quality of the fruits are affected and susceptibility to drought increases. It is also helpful to confirm the diagnosis based on visible symptoms.

Table 7: Critical level of K in fruit crops

Crops	Potassium content (%)
Avocado	0.35
Banana	2.08
Coconut	0.56
Date	1.00
Fig	0.20
Citrus	0.40
Litchi	1.11
Mango	0.25
Pine apple	0.62
Grapes	1.31
Guava	1.25
Apple	1.34
Pear	0.70
Pomegranate	1.47
Peach	1.50
Walnut	0.90
Oil palm	0.80

The critical content of K below which the deficiency is expected is furnished below.

4.2.3.5 Visible symptoms

The location of symptoms of nutrient deficiencies within the plant depends on the extent and rate of retranslocation of nutrients from old leaves to new growth. For generations, the appearance of a plant has been used by the scientists, by the lay man, by the gardeners and by farmers to diagnose the plant health. Visual symptoms are indicators of mineral deficiency symptoms.

Avocado

Potassium deficiency leads to the development of narrowed leaves of slightly subnormal size. Spring –flush leaves exhibited numerous light brown specks scattered over the leaf. As the leaf ages, the brown specks coalesced into large, irregular, reddish-brown necrotic areas between the larger veins, along the margins, or the tips. The trees are stunted, the twigs are very thin and some dieback occurrs during the winter. Its deficiency also results in interveinal chlorosis and small scattered burned spots on older leaves, part of which eventually drops out. Leaf margins are irregular. The main vein on the leaves showed a brownish colour and burning on the under side of the leaves.

Banana

Potassium shortage reduces the growth, leaves are slower to appear and smaller than normal, green at first but later developing a rapid general chlorosis, which spreads from the tip proximally. This colour is characteristically a bright (even slightly orange) yellow and is accompanied by marginal and distal scorching which proceeds rapidly through irregularly until the whole leaf is dry. Often, the dead leaf margins standing or breaks near the bottom of the lamina, rather than at the base of the petiole and this is a very characteristic symptom in the field.

Citrus

Deficiency symptoms first manifest in slowing down of growth accompanied by an excessive shedding of leave at blossom time. New shoots are weak, often poorly attached to the twigs and sometimes likely to shed. Leaves are small. Under acute conditions, there is malformation of leaves and fading of chlorophyll. Finally under very acute conditions, there is a die-back of twigs and plants are reduced in size. Other symptoms are scorched leaf tip, appearance of small, brown resinous spots peppered over various part of the leaf and gum formation in other parts of the tree. Crinkling, curling and yellowing of older leaves have also been observed. Short 's' shaped laterals with small crinkled and some what spotted leaves have also been observed. Plants deficit in K are most susceptible to cold and drought. Fruits are small, smooth and have a very thin peel. Sometimes, premature shedding of fruits also occurs. Fruit quality is also impaired.

Grapes

Potassium deficiencies begin to appear on leaves on the middle part of the shoots. Fruit's colour fades, beginning at the leaf margins, as the fading continuous, it moves into the areas between the main veins, gradually decreasing as it approaches, vein areas in the leaf centre. Marginal burning and upward or downward curling of the leaf edges follow the development of the pale green colour. In much cases, weak secondary growth may begin on the defoliated shoots in a manner similar to that which occurs when defoliation is caused by redspiders. Veins with advanced potassium deficiency carry small, tight cluster of small unevenly ripened berries.

Mango

Pronounced symptoms of K deficiency exhibited as a characteristic type of foliage necrotisis. The leaves located near the middle or slightly below the middle of the current season growth are first to be affected. From this, the injury spread to the distal and basal part of the affected twig. The scorch is usually located along the leaf margin and is preceded by dark purplish discoloration. The intensity of the scorch varies with severity of the deficiency.

Guava

Plant growth is poor and leaf size become smaller than normal and plants look sickly. The upper leaves become light green and lower leaves partially yellow in appearance with purple brown edges and deep purple small patches scattered all over the terminal end but concentrated towards the margin. The edges of the leaves become broken in more advanced stage.

Sapota

Potassium deficiency results in reduced growth, yellowing near the margin and development of dark grey spots, spreading through out the leaf surface in severe cases and poor quality of fruits.

Cashewnut

Potassium deficiency symptoms may develop in cashew seedling within 2 months after planting in the main field. Lower leaves turn yellow, starting at the apex and along margins showing necrosis symptoms spread rapidly from lower part of leaf towards the top.

Pineapple

Potassium deficiency leads to appearance of brown spots on green parts of the leaves and withering of leaf tips. The leaves are also shorter and narrower than normally nourished leaves. Initially, K deficient leaves remain green with only leaf tip drying, and necrotic spots are formed on the blistered areas which appear later on leaf surfaces. These symptoms appear first on older leaves and progress towards to younger ones. Fruits from K deficient plants are small, late maturing, low in total soluble solids and acids. Leaves of K deficient plants are markedly smaller and has necrotic tips and margins.

Pomegranate

Reduced plant growth with lanky shoots. The size and number of leaves are reduced. The leaves at the middle position on the shoots develop chlorotic symptoms on the margins with few patches on the lamina having rusty brown or blackish grey spots. Backward curling of leaves from the tips to the base also occur.

Strawberry

Trees are weak with shunted growth. Foliar symptoms started with the appearance of broad brown patches on the upper surface of leaves. The surface of leaf was puckered. There was a greater concentration of brown spot in the marginal area between the veins and they gradually enlarge and coalesce. Finally, scorching of margins took place, which was followed by premature abscission of leaves.

Coconut

Deficiency generally results in low yield due to small size nuts. The first visual symptoms appear on the leaflets as rusted coloured spots in longitudinal bands on either side of the midrib. Slight yellowing is observed which intensifies later leaving only a narrow medium band green, further narrowing down to necrosis. The rusty spots invade the whole leaflets and form large patches of irregular outline. As K is a mobile element, lower or older leaves turn yellowish and later on dries, while younger or upper leaves remain green in colour. Drooping of older leaves at the base may probably due to the inadequate supply of potassium and other nutrients. Such fronds droop down around the trunk. Under acute cases, the palm look unhealthy with dried leaves hanging all around producing very small nuts or barren nuts also. Generally, the deficiency is visible when the leaf concentration of K is less than 0.8 percent.

Oil palm

Most common symptoms are 'Orange spotting. Pale yellow spots develop along the leaflets of older leaves. As the deficiency becomes more severe, spots become large and orange in colour. At the end, the entire leaf gets a rusty brown colour and dies. Deficient palms yield very few and smaller fruit branches.

Apple

Potassium deficiency is exhibited in all parts of the plants. Leaves become bluish green and wavy. Marginal chlorosis is found in middle leaves. The dark to reddish brown necrotic lesions at the leaf margin (scorching) spread towards the middle of the leaves. The leaves remain on the tree for a considerable time after drying. The fruits are small, dull and pale. Potassium deficiency also leads to thick skin of the fruit and poor colour development. The flesh of the fruit is sour and poor in sugar.

Pears

Leaves are folded or uprolled and have a pale yellowish – green to silvery – brown appearance. Brown to dark brown necrotic lesions develop along their margins. Tree growth is poor.

Peach

The middle and older leaves are yellowish green, wavy, forward curled and bent downwards like a sickle. The scorched tips and margins of the leaves are reddish brown to purple, tear and have a ragged appearance. The twigs are thin and blossom and fruits are poor in growth. The shoot tips die, and the few fruits are small and mishapen.

Plum

Brown, dark brown and purple necrotic lesions appear along the cracked margin. The tips of shoots die, and the fruits are small, remain red and unpalatable.

Cherry

The bluish green leaves of the cherry tree fold or roll upwards parallel to the midrib. This is followed by the appearance of brown marginal necrotic lesions.

Red currants

Leaves are bluish green with slight chlorosis and brown marginal necrotic lesions. The edges are unrolled and the leaf blades bend downwards. Similar signs of scorching are shown by black currants, but they are with reddish purplish colour. The berries ripen unevenly and lack sweetness.

Litchi

Potassium deficiency in litchi cause considerable stunting of trees. The leaves were small and soft in texture. There was scorching of leaf margins starting from the tip and progressing towards the base of the leaflets. K deficiency also causes necrotic areas at the apices of the leaflets which gradually progress along the margins towards the base of the leaflets. There was heavy defoliation.

Peach

The K deficiency in peach is expressed by small leaves near the ends of terminals and on lateral shoots. These leaves are usually long in relation to width, lighter green than the older leaves and showed a tendency to fold and roll. In acute K deficiency, the older leaves show crinkling and puckering of the tissue along the midrib. These spots later develop into necrotic areas becoming brown and the injured tissue is dropping out of the leaf blade. Thus, giving a wavy margin and shot-hole effect to the leaf. The twigs were slender with scant fruit buds on them.

Walnut

Potassium deficient leaves begin showing symptoms in early to mid-summer. Leaves become pale and later the edges of the leaves fold upward and curl in. The exposed undersides of the curled part of the leaflet develop a greyish cast (bronzing). Generally, most of the affected leaves are along the middle of the shoots. Leaf size is reduced as well as shoot growth. Invariably, overall nut size also is reduced.

Papaya

Potassium deficiency in papaya is expressed in the form of tip and marginal burning on leaves, retarded stem growth and flowering suffers adversely.

Strawberry

The leaves turn bluish green besides the midrib. The edges of older leaves are reddish brown to purplish, later turning brown, and drying. Discoloration and necrosis advances until only a small green triangle remains at the base of the leaf. Potassium deficiency also causes a dark brown discolouration of the base of the leaf which spreads into the leaf stalks. Light brown necrotic lesions then appear in the discoloured zone. Fruit development is poor and the fruits are less palatable.

Raspberries

The potash deficient plants have dull green leaves and their margins tend to curl backward. The foliage shows brown marginal scorch which may extend interveinally. The growth of the crops is stunted.

Blueberry

Leaves show interveinal chlorosis; older leaves have necrotic spots and marginal scorch; terminal growing point aborts.

4.2.3.6 Remedial measures

Potassium is supplied to crops as straight fertilizers or in the from of compounds. Generally K fertilizers are usually applied broadcast. In soils with low potassium, row or band placement of K is more often effective than broadcast dressings.

Table 8: Recommended potash for fruits crops

Crop	(Kg K_2O /tree) Adult tree.
Apple	1.000
Banana	0.330
Mango	1.500
Sweet orange	0.300
Mandarin orange	0.400
Grapes	1.200
Guava	1.000
Sapota	1.500
Papaya	0.200
Pomegranate	1.200
Jack	0.500
Ber	0.500
Plum	1.000
Peach	1.000
Cherry	0.700
Cashewnut	0.125
Walnut	60Kg/ha.

4.2.3.7 For effective management of K

1. Generally plants require more K during early age of crop growth. Hence, adequate K should be made available at vegetative stage by basal application of K.
2. Gypsum may be applied to increase the K availability in perennial crops.
3. Besides the sufficient K in soils, K fertilizers should be applied in soils

particularly during the cold season to ensure more availability and easy uptake by plants.

4. Sandy, peat and muck soils require more K due to heavy leaching of K.
5. Placement of K fertilizers in bands reduce the contact between soil particles and fertilizers droplet and reduces K fraction and hence increases K availability.
6. Compost contains very little amount of K and hence compost can not meet out the whole requirement of K and therefore compost should be combined with KCl.
7. In compact soils, either the soil should be loosen or potassic fertilizers should be applied especially in bands.
8. Irrigation increases K diffusion and hence increases K supply to plant roots
9. For calcareous and over limed soils, K fertilizers application should be encouraged even though the soil contain enough K.
10. Potassium fertilizer should be applied in splits and broadcast in acid soils.
11. Application of K fertilizers in drought or dry soils will not produce desired response due to reduced K accessibility to plant roots.
12. In general, split application of KCl is more effective than its heavy application as single dose.
13. In flooding condition, $K_2 SO_4$ should not be added, otherwise, it will produce H_2S and make unavailability of Zn, Cu, Fe and Mn.
14. For coconut, application of NaCl to some extent act as a substitute for K.
15. Application of K fertilizers in bands in close proximity to seed may cause damages to seedlings due to high K concentration. Hence, potassium fertilizers in bands should be applied horizontally about 5 cm away for the seeds or about 2.5 cm below the seeds so that roots will develop in zone free from concentrated fertilizers solutions.

4.2.4 Calcium

Calcium is one of the secondary elements because its requirement is being less than that of primary nutrients. The role of Ca as soil conditioner has been recognized more widely than as an essential plant nutrient. The deficiency of Ca is evident in acid soils as well as in alkali soils. Calcium deficiency has rarely been reported in normal soil owing to its large quantities available in soil solution. However, due to the introduction of high yielding varities /hybrids, improved farming practices and intensive cropping resulted in enormous removal of Ca from the soil. In general, crops remove Ca in the range of 11 to 196 Kg Ca/ha. It is also reported that there is as much as higher removal of 300 Kg/ha in one-year growth of Cavandish banana. The importance of Ca in fruit production is not only felt in increased fruit yield with quality, but also on the shelf life of fruits.

4.2.4.1 Calcium in soils

Calcium has been known as an essential element for plants for a long time. Sprengle (1839) was the first to show that Ca is essential for plant growth. Calcium is the fifth most abundant element in the earth and constitutes 3.64% of the earth's crust. Calcium in soils originated in rocks and minerals from which the soils were formed. They include calcite, augite, dolomite, anorthite and biotite. Calcium content of the soil varies considerably. Sandy soils of humid regions contain very low amounts of Ca^{++} while in non-calcareous soils of humid temperate regions, Ca^{++} ranges between 0.7% and 1.5%. In high weathered soils of humid tropics, Ca content is very low (0.1 to 0.3%). Ca in calcereous soils ranges between less than 1% to 25%. Calcium content more than 3% indicates the presence of Ca carbonate. The Ca content of arid region soils is generally high regardless of texture as a result of low rainfall and little leaching. Calcium content of humid soils are low due to the removal of Ca by excessive leaching.

4.2.4.2 Factors affecting Ca availability

a) Easily leached sandy soils contain low Ca. Soils with less than 1.5me/100g of Ca is considered as deficient in Ca.

b) Low soil pH will impede Ca^{++} uptake ($pH < 5.5$)

c) The amount of Ca^{++} absorbed by plant decreases as the % of Ca^{++} saturation decreases in proportion to the total CEC.

d) Crops respond to Ca application when the Ca^{++} saturation falls below 25%.

e) The Ca availability also depends upon the type of clay. 2:1 clay requires higher Ca^{++} saturation than 1:1 clay.

f) Ca availability to the plant is also influenced by ratios between Ca^{++} and other cations in solution. A Ca^{++} /total cation ratio of 0.10 to 0.15 is desirable for an adequate Ca^{++} supply to most crops.

g) Application of nitrate nitrogen stimulates the Ca absorption; whereas Ammoniacal nitrogen, potassium, magnesium and manganese suppress the uptake of Ca^{++}.

h) High humidity combined with high temperature may also induce Ca deficiency.

i) Irregular water availability coupled with reduced transpiration may also cause reduced Ca^{++} uptake.

j) High K content in leaf influences Ca^{++} distribution in plants and can thus induce deficiency.

k) Low O_2 availability in soil due to compaction or loss of soil structure may also reduce the uptake of Ca^{++}.

l) Wet soil, including wetness due to poor drainage reduces Ca uptake.

m) Excessive application of Boron also reduces Ca uptake.

n) Too much of shade, inhibited transpiration, reduced root growth may also lead to reduced Ca uptake.

o) Prolonged day period with high temperature also reduces the Ca uptake.

4.2.4.3. Calcium in plants

Calcium is absorbed by plants as Ca^{++} ion from the soil solution and is supplied to the root surface by mass flow and root interception. The absorbed Ca^{++} is transported to the above ground plant parts with transpiration flow through the xylem. Generally, the Ca^{++} concentration of the soil solution is about 10 times higher than that of K^+. The uptake rate of Ca^{++}, however, is usually less than that of K^+. Capacity of plants for Ca^{++} uptake is limited because it can be absorbed only in young root tips where the cell walls of the endodermis is unsuberized. Calcium occurs in plant tissues as free Ca^{++}. As Ca^{++} adsorbed to indiffusible ions such as carboxylic, phosphorylic and phenolic hydroxyl groups. Calcium is also present as oxalates and carbonates in cell vacuoles.

Crops with small root system may face problems of inadequate uptake of Ca^{++} than those with highly developed root system. Although the Ca^{++} in soil solution is higher than the plant's requirements, it 's uptake by the plant is limited because the Ca^{++} uptake is genetically controlled. Crop species and their varieties differ greatly in their Ca content and ranges between 0.2% and 5.0% (dry weight basis) in leaf tissue. A Ca content of 0.5% is considered as adequate for plant growth. The Ca uptake of (monocotyledons) cereals and other grasses is much lower than that of dicotyledons, particularly in herbaceous plants and legumes, whose Ca needs are five times as high as those of grasses. Tomato contains 10-15 times higher Ca^{++} than those of wheat. The higher Ca requirement of dicots are related to the higher CEC in the roots and other plant parts. Since Ca is an immobile element, leaves contain higher Ca than the stem and fruit or seeds. Plants are not able to utilize Ca^{++} from older leaves for the growth of meristematic tissue, due to its immobile nature of Ca^{++}.

Based on the adaptation of the plants to varying pH and Ca status, the plants are classified into two groups, viz., Calcicoles and Calcifuges. Calcifuges are typical of the flora observed on calcareous soils where as the Calcifuges are grown on acid soils which are poor in Ca. In fact, many of the Calcicoles species contain higher levels of intracellular Ca and high concentrations of malate; whereas the Calcifuges are normally low in soluble Ca. Plants of calcicoles-calcifuges differ in their ability to utilise Fe. It was noted that plant species originating from acid soils are more susceptible than those from calcareous sites.

4.2.4.4. Calcium in physiology

Calcium, as a constituent of phytin which combines with phosphorus to form Ca phosphate and act as buffers when pH values change and can also be regarded as a reserve phosphorus supply for the plant. Calcium is essential for cell division and cell elongation. It is also needed for chromosomal and mytochondrial stability. Calcium ions are important for the stability of the structure of protoplasm. Calcium is deposited on the cell wall as Ca carbonate incrustation, and is found in plants in the form of Ca oxalate. Calcium as Ca pectate forms the constituent of the middle lamella of cell walls.

Calcium is an important nutrition for root development and functioning. Calcium is required for membrane permeability and cell integrity. Calcium is important for the regulations of phytohormones. Ca is needed for the reduction of nitrite. Ca activates as many as 20 enzymes in metabolism. However, the action of Ca is non-specific and the Ca^{++} ions can be replaced by Mg^{++} and Mn^{++}. Out of these, the role of Ca in amylase and ATPase are well known. Lack of Ca^{++} ions in legumes prevent the development of the nodule bacteria, thus Ca^{++} is involved in N_2 fixation. Calcium is important for protein synthesis and carbohydrate transformations. Calcium also retards senescence. Germination and growth of pollen tubes are induced by Ca^{++}. Calcium also plays an important role in increasing the shelf life of post-harvest fruits by regulating respiration. Low content of Ca^{++} in fruits accelerates ripening by increasing respiration and reducing firmness and thus Ca deficiency results in reduction of shelf life of fruits. High concentration of Ca^{++} in plants inhibit the toxic effect due to the uptake of heavy metals by plants. It is also involved in the improvement of fruit quality in terms of higher juice content with higher soluble solids. Ca also seems to increase the tolerance of salinity.

4.2.4.5.Diagnosis of Ca deficiency

The expression of Ca deficiency is closely related to the specific behaviour of Ca^{++} ions in plants viz., migration with the transpiration flux, accumulation in older leaves , immobility and the functions of calcium in plants. Because of these factors, the symptoms manifest themselves first on young tissues and organs where the differentiation is still taking place. Plant growth is therefore inhibited, and the plants have a bushy appearance. The youngest leaves which are first affected are usually small and mishappen, having a ladle-like appearance with upcurled tips and margins. A typical feature of calcium deficiency is that the veins of completely necrotic leaves are always darker than intercostal regions. The deformation and crinkling of young leaves with up curled and sometimes turn necrotic margins and the hook, claw or hood like curling of the tips and margins are poorly developed. The leaf stalk may also be curved backwards. The leaf margin sometimes roll up inwards. The leaves die from tip to the base, as also do the stalks (which turns a dark, blackish colour in the process), the apical meristem tissue of the terminal buds and the roots. If calcium alone is lacking, the older leaves often have enlarged blades and somewhat darker green colour. Calcium deficiency symptoms has been rarely observed in fruit crops under field conditions. However, the problems relating to the physiological disorders of post-harvested fruits due to inadequate Ca has become ever increasing.

Plant analysis, as a diagnostic technique , not only gives a clear cut picture about the nature and extent of hidden hunger nutrient, but also used to verify the visible deficiency symptoms. The plant parts to be sampled for analysis varies among the crops. In general, mid shoot of current growth prior to flowering can be collected. The critical content of Ca below which the deficiency is expected is given below for a wide range of fruit crops.

Table 9: Critical Ca contents in fruit crops

Crops	Calcium content (%)
Banana	0.54
Apple	0.58
Grapes	0.73
Papaya	1.81
Mango	1.51
Citrus	1.60
Guava	0.67
Cherry	0.80
Almond	1.00
Apricot	1.00
Pears	0.80
Avocado	0.50
Oil palm	0.25
Date palm	0.22
Cacao	0.30
Cashew nut	0.24
Pineapple	0.10
Passion fruit	1.00

Critical values for Ca vary considerably among crop species, lowest for the grain crops and highest for some vegetable and more fruit crops. Benton Jones (1998) suggested that the critical content based on the total Ca in plants may not be reliable since most of the Ca in plants accumulates as calcium oxalate crystals. Therefore, extractable Ca (in 2% acetic acid) may be a better indicator of sufficiency. The critical Ca concentration for soluble Ca is around 800 ppm.

4.2.4.6 Deficiency symptoms

The deficiency symptoms of Ca in fruit crops have been documented mainly based on the experiments conducted under controlled conditions. These visible symptoms may serve as a ready reckoner for the fruit growers to identify the deficiency of Ca.

Apple

Calcium deficiency has been observed on the roots and the fruits. Root growth is restricted and the root-tips become brown. Large spots coupled with chlorosis may develop in the centre of the leaves. Leaves on the tips of the shoots may turn yellowish and be curled upwards. Tip of the shoots may die. Root growth is restricted and results in loss of vigour. Perhaps, the loss of vigour by old trees and failure of young trees to live and grow may sometimes be the effect of Ca deficiency. Apples with low Ca result in number of physiological disorders viz., bitter pit, internal breakdown, fruit cracking, cork and sunscald.

Banana

The deficiency has been characterized by a narrow marginal chlorosis of the older leaves which leads by way of necrotic flecking to a general but strictly marginal chlorosis.

Citrus

Calcium deficiency in citrus has been characterized by fading of chlorophyll along the leaf margins and between the main veins, appearance of small necrotic spots in the faded areas, and premature leaf fall followed by additional flushes which are also shed. Twigs die back and development of weak shoots also soon die. Stunted growth of trees and formation of thick hard peel in fruits occur. To begin with, leaves are dark green, but turn yellow later with discoloured edges.

Cacao

Newly formed leaves show necrotic areas between veins, symmetrical on each side of mid-rib. Premature leaf shedding also occur.

Cashew

Cupping of young leaves with marginal scorch and death of growing points are the deficiency symptoms of Ca.

Coconut

The deficiency has been observed in seedlings and found that the seedlings are tall and slender with spare leaves taking on a "tucked in" appearance. The petioles of the leaves are of a deep yellow or orange colour, orange blotches along the mid-rib occurring frequently. Leaflets show angular patches of an orange-yellow colour, they then wither and finally die.

Cherries

Light brown to yellow markings occur on leaves. Leaves may become tattered with numerous holes and shoot growth is minimal. Fruits with low Ca are sensitive to rain induced crackings at harvest time.

Grapevine

Youngest leaves are smaller than normal. The margins and intercoastal surfaces are chlorotic, and necrotic spots (pin-head size) appear along the leaf margins. Leaves often curl inwards, and the tips curl downwards, giving the leaves a hook-like or claw like appearance. The vegetative points and shoots die.

Litchi

Calcium deficient plants produce smaller leaflets. Necrotic leaflets give the appearance of scalloping of the margins. There will be heavy leaf fall.

Mango

Plants under Ca deficiency are smaller and the leaves show yellowish green colour. Calcium deficient plants look somewhat similar to nitrogen deficient plants, but in the case of Ca deficiency, the apical leaves show yellowing first; whereas in nitrogen deficiency, the basal leaves are affected first. In case of severe Ca deficiency, the older leaves show brownish scorching all along the margins except small areas near the tip and the base. Such leaves turn yellow within a week.

Peach

Young leaves of terminal growth die back from tips and margins or along midrib. Following extreme deficiency, twigs may die back; most leaves are shed; roots are short, bulbous and die back after making little linear growth and new roots emerge profusely behind dead root tips.

Pomegranate

Apical growth ceases, foliage becomes sparse and hooking of leaf tips occur. Terminal leaves become chlorotic and white patches develop in advanced stages.

Pineapple

Leaves are pale green, with some yellow mottling, die back occurs at the tips of the new leaves; as the shortage becomes more severe, the leaf begins to blister and develops a reddish colour at the base; fruiting is premature and the fruits show areas of internal discoloration containing gelatinous materials.

Blueberry

Leaves become chlorotic ; necrortic in areas with marginal scorch.

Raspberry

The young leaves of the bushes are chlorotic and exhibit necrotic lesions on either side of the middle rib.

Strawberries

Young leaves may die or only leaflet or portion may be affected; appearance of red-brown colour around the base of leaflet; roots show injury before tips are affected, drying back from tips followed by development of new rootlets behind the dead portion; entire mass of root system is affected.

4.2.4.7 Physiological disorders

Fruits with low concentration of Ca are highly susceptible to a number of post-harvest and storage disorders. Calcium deficient apples frequently suffer from "bitter pit", "internal breakdown", "cork", "sunscald", and "scald". Calcium deficit pears associated with cork spots of "Anjon and Black" end in 'Bartlet" pear. "Impact damage" and "splitting of fruits" are the disorders in cherry due to lack of Ca. "Fruit splitting" in general among fruits is due to Ca deficiency. "Soft nose", a physiological disorder characterised by a breakdown of the flesh at the apex of the mango fruit, is thought to be caused by low Ca content in fruits. Ca deficient grapes dispalys "necrosis of pedicel". At full development of stage of berries, the pedicel becomes black-brown and later turn black and subsequently berries dry. There is a shrinkage in pedicel girth. Drying starts from the tip of the bunch in Thompson seedless. The 'uneven ripening' of sapota is also due to low Ca in fruits.

4.2.4.8. Correction of Ca deficiency

Calcium deficiency can be easily managed by liming the soil if the soil is acidic, that is, with pH value below 6.0 due to leaching of Ca. Ordinary limestone can be used as liming material if the soil is not deficient in Magnesium. In such conditions, addition of single superphosphate or rock phosphate will supply adequate Ca to the plants. In general, Ca is applied in the ranges between 3 to 4 t CaO or 4 to 6 t $CaCO_3$ /ha with an interval of 3 or 5 year cycle. The quantities applied not only depend on soil pH, but also on the content of H^+ adsorbed to soil colloids. If the soil is slightly acidic, neutral, or alkaline, gypsum will be the appropriate choice of Ca. Applicaton of Ca ammonium nitrate and Ca nitrate (during winter) not only supply N, but also meet Ca requirement. Foliar sprays of Ca enhanced the keeping quality of fruits. Pre-harvest sprays of 1.0 % calcium nitrate on peach and 0.6% calcium chloride on Muscat grapes prolonged the keeping qualities of fruits owing to delayed onset of senescence.

4.2.5 Magnesium

Magnesium is considered to be an important secondary nutrient in the whole of the plant kingdom. Eventhough plants need small amounts of Mg, its active role as central atom of chlorophyll molecule makes life impossible without it. It's essentiality for plant growth was established in 1906 by Willstatter. The importance of Mg in fruit production is largely being felt in recent years due to the introduction of high yielding fruit crops.

4.2.5.1 Magnesium in soils

Magnesium constitutes 1.93% of the earth's crust and its content ranges from 0.1% in coarse sandy soils in humid regions to 4% in fine structured, arid or semi-arid soils. Magnesium originates in the soil from the weathering of rocks of the parent materials viz., basalt, peridotite and dolomite. The exchangeable Mg^{++} occurs in between 4 and 20% of the CEC of soils. Under average conditions, exchangeable Mg is about 5% of the soil's cation exchange capacity. Magnesium occurs in soils in a slowly available form which is in equilibrium with exchangeable Mg^{++}.

4.2.5.2 Magnesium in plants

Magnesium is absorbed by roots of the plant as Mg^{++} from the soil solution by mass flow and diffusion. Mg is a very mobile element in the phloem and can be translocated from older to younger leaves. The plant tissue contains Mg^{++} in association with mobile anions such as malate and citrate and with non diffusible anion such as oxalate and pectate. Crops differ in their uptake and response to Mg. The root crops and legumes usually require Mg double the quantity of cereals and grasses. Magnesium contents are even higher in oil yielding crops. Crops viz., potatoes, oilpalm, tobacco, citrus and sugarbeet are highly responsive. The Mg content of plant ranges from about 0.1% to 0.5%. The Mg present in the chlorophyll molecule is only about 15-20% of the total Mg content of the plant. The average uptake of Mg in crops range between 10-25 Kg/ha/ year.

4.2.5.3 Magnesium in physiology

Magnesium plays an important role in the construction of the chlorophyll molecules as it occupies the central position of the chlorophyll molecule. It also favours the synthesis of carotene and xanthophyll. It activates nearly 300 enzymes mainly as cofactors. Magnesium acts as a cementing material between ATP or ADP and enzyme molecules in the phosphorylation reaction. Enzymes viz., dehydrogenase and enolases are also activated by Mg^{++}. A note worthy and very important function of Mg^{++} is synthesis of carbohydrate through the activation of the world's most abundant enzyme Ribulose Bisphosphate carboxylase. Magnesium is involved in the stabilization of ribosomal particle in the configuration necessary for protein synthesis. Magnesium is also needed for synthesis of oil in plants. It also activates the formation of polypeptide chains from amino acids. Magnesium is also needed for phytohormone balance and the reduction of nitrate. Recently, Mg^{++} is also thought to be involved in the promotion of germnation of pollen. Mg influences the strength of cell walls and permeability of membranes. It also increases drought resistance in plants.

4.2.5.4 Conducive factors for Mg deficiency

1. Magnesium deficiency is common in plants growing on coarse textured acid soils having a sandy loam or sand surface textured with subsoil as coarse or coarser than the ploughed layer.
2. A very high dose of potassium fertilizers, above that required for a maximum yield on a sandly soil induces Mg deficiency. Generally, the recommended K/Mg ratios are $<^5/_1$ for field crops, $^3/_1$ for vegetables and sugarbeets and $^2/_1$ for fruit and green house crops.
3. Magnesium deficiency also occurs in soils with high ratios of exchangeable Ca/Mg (This ratio should not exceed 10/1 to 15/1).
4. Magnesium deficiency is also created when higher dose of ammonium sulphate is added to low exchangeable Mg^{++} soils.
5. Acid soils with pH 5.0 and below are liable to exhibit Mg deficiency.
6. Soils with less than 3-4 mg/100 g of Mg is considered as critical content of magnesium and this values differs according to the soil texture.

4.2.5.5 Diagnostic technique

Since it is mobile in plant tissue and deficiency always begins in the older leaves and then move to the younger leaves. The very characteristics symptoms of magnesium deficiency which appear first on the older leaves, are closely linked with the decomposition of chlorophyll. Since Mg^{2+} ions are withdrawn more slowly than nitrogen and Mg^{2+} ions are repeatedly pressed into metabolic services again as they migrate towards the veins, a characteristic more or less broad green margin remains on either side of the main vein (vein banding). In rapidly growing plant, Mg deficiency may also be observed first in younger leaves if Mg withdrawal from the older leaves is too slow. In dicotyledons, the intercostal chlorotic leisons typical of Mg deficiency, which are usually bright yellow, although, a violet tinch is observed in some species. Leaves from Mg deficient plant are shed prematurely, giving rise in the case of fruit trees to 'brush disease' in which long shoots are denuded except for a few leaves forming a rosette at the tip.

Plant analysis has been used not only to identify the deficiency element but also to measure the nutrient content in plant tissue. In general, fully matured leaf from the current season shoot growth is taken as sample for analysis and may slightly vary depending upon the crops. The critical content of Mg in fruit crops are given below.

Table 10: Critical content of Mg in fruit crops

Crop	Critical level (%) (less than)
Cocoa	0.20
Coconut	0.30
Citrus	0.20
Peach	0.25
Grapes	0.20
Mango	0.16
Oilplam	0.20
Avocado	0.15
Walnut	0.30

4.2.5.6 Visual symptoms

Apple

Leaves show whitening or fading known as chlorosis. The leaves may show an inter-veinal scorch or burn which may extend to their margins in old trees. Chlorosis and scorch are likely to appear suddenly in mid-summer and rapidly progress from the base to the tip. The leaves and the lower part of the shoot drop early, but the tip leaves of the shoot remain attached till normal leaf fall. The pre-harvest drops in apple is high. The fruit drops earlier than normal maturity. The size, yield and quality of fruit is also marginally affected.

Banana

The older leaves look pale in colour and develop a general necrotic flecking, the flecks later acquiring yellow haloes; the leaf margins are little affected. Sometimes, purple blotches appear on the petiole and in the severest cases, stunting, choking and breaking away of the leaf sheath takes place.

Avocado

Triangular green area appear towards midrib of leaf; marginal chlorosis extends interveinally toward the midrib. General chlorosis finally develops, with veins remaining green, followed by dead lesions scattered over the entire blade.

Citrus

In the early stages, yellow areas develop between the large veins and on both sides of the midrib. These gradually enlarge and coalesce, producing general yellowing first in the base of leaf and later in the entire leaf. Sometimes, prominent yellowing on both sides of the midrib extends from the base to the apex of the leaf leaving an inverted 'V' shaped green colour along the basal portion of the midrib tapering towards the leaf tip.

Coconut

The first deficiency symptom is the inter-vascular yellowing of the older leaves. The petiole of the leaf and the midribs of the leaflets also remain green. In the early stages, a small margin on either side of the midrib of the leaflets also remain green. Gradually, the yellowing proceeds to younger leaves and the older leaves turn to yellow orange and even to orange colour. The younger leaves remain green. The leaves exposed to the sun are easily prone to yellowing. Older leaves may assume a bronzed appearance and dry.

Cashew

Interveinal chlorosis and yellowing of lower leaves are the characteristic visible symptoms of Mg deficiency. The symptoms spread rapidly from the lower leaves towards the upper leaves.

Cherry

The interveinal parts of the older leaves become chlorotic and turns to orange or reddish tinge. Later, oval brown to dark purplish brown intercostal necrotic lesions develop in a herringbone pattern. Leaves are shed prematurely.

Cocoa

Growth is very poor, stems are spindly and there is very little rootlet development. The leaves become small in shape, uniformly yellow-green with light green veins. The leaves turn yellow, margins and tips develop scorching.

Grapes

Magnesium deficiency can be visualised in two forms. Prior to flowering, interveinal and marginal leaf necrosis dominates. During summer leaf fall, interveinal yellowing or reddening and the development of brown-green patches are the major symptoms. As discoloration increases, it proceeds from the margin of the leaf in a wedge towards the petiole attachment point.

Gooseberries

The older leaves of gooseberries exhibit a very broad red or brownish to yellowish red belt, which gradually turns cream, leaving a pale green centre at the leaf stalk. The leaves are shed prematurely.

Litchi

The deficiency results in smaller leaflets. Tiny necrotic areas appear parallel to and in almost equidistant from the midveins of the leaflets. The necrotic areas become bigger until some of them became continuous. The plants fail to bloom and there is heavy defoliation.

Mango

The trees become stunted in growth and pale in appearance. On the developing leaves, yellowish white areas between the principal veins on both sides of the midrib run parallel from the base to the apex followed by chlorosis and then start spreading outwards towards the margin.

Blueberry

Basal leaves show marginal and interveinal chlorosis, followed by reddening and necrosis, younger leaves may be so affected later.

Oil palm

Deficiency causes orange yellow patches of discoloration which occurr in areas between veins on either side of the midrib, leaflets which are exposed to sunlight show yellowing or orange discoloration. In advanced stages, dark necrotic spots and scorching from the tip are observed and the leaflets dried up. In the case of adult palms, these symptoms are most striking with bright orange yellow colour which is designated as 'orange frond'.

Pomegranate

Interveinal chlorosis of basal leaves with green midrib and veins occur. Necrosis also occurs.

Amla

Older leaves turn slightly pale yellow and show broad red marginal bands which increases in width. Red colour fades to cream.

Pear

Appearance of oval intercostal necrotic leisons arranged in a herringbone pattern along the middle rib appear in the region of the margins and the margin themselves remain green, the leaves are shed prematurely.

Peaches

The middle leaves are the most affected and exhibit slight, but easily discernible, intercostal chlorosis. Dark green 'water spot' with purplish-red borders around the dead zone appear on the older leaves of annual shoots. The blotches turn a greyish, green-white or light brown to brown. The leaves fall prematurely.

Pineapple

Insufficient magnesium content results in pale greenish yellow leaves with development of yellow spots on the margins at later stages and reduction in fruit firmness.

Plum

Older leaves become chlorotic between veins, followed by necrosis and usually margins are first affected.

Raspberries

The deficient leaves develop reddish tinge and the veins are banded. The leaves are shed from the bottom to the tip of the plant.

Strawberries

Small, reddish-purple blotches forming a linear pattern appear between the veins. The blotches merge to form belts between the green veins. Finally, the whole leaf turns purplish red, except for a green band . Yellowish green chlorotic blotches sometimes appear near the main nerve before the leaves turn reddish-purple.

Walnut

Basal lower leaves may show marginal interveinal yellowing and dying with a tendency to drop in midseason or early fall, although there may be an early leaf drop or burning if the deficiency is acute.

4.2.5.7 Remedial measures

Magnesium deficiency can be rectified by supplying the sources of Mg fertilizers. They can be applied through soil or foliar or by seed treatment.

$MgSO_4$, $MgCl_2$, $Mg(NO_3)_2$ and synthetic chelates are well suited for foliar sprays. Domolitic limestone is recommended for acid soils. The low pH of the acid soil promotes the availability of Mg to the plant. Magnesium in the form of sulphates are commonly available and largely being used. Epsom salts are more soluble than the Kieserite and hence, Epsom salts are found to be more effective. For soil application, sandly soils require more Mg than the soils with more clay. The quantity of Mg fertilizer depends upon the nature of the crop and its availability in soil. Farmyard manure also contains considerable amount of Mg. Foliar sprays of 1 per cent epsom salt applied 3-4 times during the early stages of crop growth is very effective in controlling the deficiency. Epsom salt is compatible with most of the pesticides. It has been also found that soaking the seeds in 2% epsom salt also avoids the occurrence of Mg deficiency. Persistence of severe Mg deficiency over a long period of time can also be managed by judicious application of potassium fertilizers.

4.2.6 Sulphur

Sulphur is one of the most important secondary nutrients required for plant growth. It is needed in similar quantities as Phosphorus or Magnesium. In 1911, W.H.Peterson discovered the essentiality of sulphur for plant growth. The continuous exploitation of soil available S, use of S free high analysis N fertilizers, soil erosion leading to loss of the top soil, introduction of high yielding varieties and inadequate supply of organic manure, have led to recognition of S as a limiting element in agriculture.

4.2.6.1 Sulphur in soils

Sulphur is the thirteenth abundant element in the earth's crust, average ranging between 0.06 and 0.10 percent. Normal agricultural soils have 50-300 ppm in the top soil. Sulphur in soils have been derived from weathering of rocks and soils viz., gypsum, barite, epsomite, mirabilite. Sulphur occurs in soils in both inorganic and organic forms. Soil S is highest in top soil and decreases with increasing depth of the soil. Normally in non-calcerous soils, organic S provides the major source of S to the plant. The inorganic forms of S in the soil arises from the decomposition of organic S by microbes and additions through rain, irrigation, water and fertilizers. Sulphur in the form of adsorbed SO_4^{2-} represents the readily available fraction to the plants. It is absorbed by roots by mass flow and diffusion processes. The C:N:S ratio in soil organic matter is approximately 125:10:1.2. The soils with less than 10ppm available S is considered as critical level. However, the critical level may range between 8 and 30 ppm depending upon the method of extraction, soil type and the nature of the crop.

4.2.6.2 Factors influencing S availability

Soils of the humid temperate and tropical regions contain more S than the soils of arid zone. In general, soils with more organic matter contain higher S than the soils with low organic matter. Absorption of sulphate is higher at low pH than the higher pH. Increased use of P leads to higher leaching losses of sulphate ion. This is because the phosphate ions compete with sulphate ions for the absorption sites of the soil. Hence, higher use of phosphatic fertilizers may induce S deficiency. Fine textured soils contain generally more S than the coarse textured soils and hence S deficiency is likely in coarse textured soils. Intensive cultivation of pulses and oil seeds exhaust more S in the soils and thus S is depleted. Use of S free fertilizers and minimum or nil use of S containing pesticides, are some of the factors for S deficiency.

4.2.6.3 Sulphur in plants

Plants aborb S in the form of SO_4^{2-} ions by the roots and transporated within the plant in the same form and then reduced to SH (Sulfhydryl-group) and incorporated into organic forms. Sulphate is mainly translocated in an upward direction and the capability of higher plants to move S in a downward direction is relatively low. Sulphur content in the plant ranges between 0.1% and 0.5% with an average of 0.3%. Sulphur content among the plant parts varies considerably. The economic part of the plant (grain, fruits) accumulates higher S content than the non-economic parts. The sulphur content of the older leaves is higher than younger leaves. Generally, the members of cruciferae legumes, oil seeds, vegetable crops require relatively higher S than the cereals and potatoes . Sulphur removal from the soil by the crops ranges between 5kg and 80kg S/ha. Sulphur uptake is generally 9-15% of the N uptake.

4.2.6.4 Sulphur in crop physiology

The quantities of S absorbed in relation to yield and its involvement in plant metabolism make it to consider S as the fourth major element after N,P, and K. The absorbed SO_4^{2-} ions in the plant cell is reduced to SH groups and are incorporated for the synthesis of amino acids viz., methionine (21%S), cysteine (27%S) and cysteine (26%S). In plant metabolism, cysteine and cystine form a redox system. These amino acids are indispensable for protein synthesis. Nearly 50% of the total sulphur content of protein being reportedly contained in the protein.

Sulphur is a structural constituent of ferredoxin, biotin, lipoic acid and thiamine. These CoA enzymes are essential for photosynthesis and respiration. Sulphur is important for the synthesis of lignin and sterols. Methionine serves as precursor for the biosynthesis of ethylene. Glutothione, a sulphur containing compound available in plenty in plant cells, plays an important role in detoxification of unfavorable compounds for plant growth. Glucosinolates, a sulphur rich secondary metabolic product is responsible for the characteristic taste and flavour.

Sulphur is essential for the chlorophyll synthesis owing to its role in protein synthesis. Chloroplasts are also rich in organically bound Sulphur. N/S ratio of protein varies slightly in plants (30/1 to 40/1). Sulphur is important for synthesis of oil in plants. Sulphur containing enzymes are involved in the reduction of nitrate nitrogen to proteins and also in legumes - rhizobium symbiotic nitrogen fixation. Sulphur is the main constituent of diallyldisulphide, which imparts characteristic flavour and causes tears in eyes due to onion and garlic. Sulphur influences the uptake of phosphorus in plants. Sulphur in the form of SH group provides cellular resistance to dehydration caused by drought, heat and frost damages and thereby S imparts resistance in plants to the above stresses.

4.2.6.5 Diagnosis of sulphur deficiencies

Similar to other nutrients, Sulphur deficiency has been detected through soil testing, plant analysis and visual symptoms. Crops deficient in S are therefore smaller than normally developed plants. If the deficiency occurs at an early stage, they are stunted and have a rigid erect appearance like that induced by lack of nitrogen. The leaves are smaller and narrower than usual and the stems are of longitudinal growth and thin stem growth is affected more severely than root growth. The younger leaves of plants lacking sulphur turn various shades ranging from pale green to completely yellow, including the veins, which sometimes take on a reddish or purplish hue if the deficiency is severe. Chlorosis in some cases is restricted to narrow bands on the side of the veins which later broaden until they cover the whole leaf. The symptom of S deficiencies are more variable than those of nitrogen deficiency.

Plant analysis is the more reliable method, although soil testing does identify potentially S deficient soils. A number of extracting solutions have been used to extract the soluble and adsorbed SO_4^{2-} (hot water, $CaCl_2$ and phosphate). The critical S content of plant tissue have been standardised. The critical content for some of the crops are given below.

Table 11: Critical content of S in fruit crops

Crop	Critical level(per cent)
Cocoa	0.20
Coconut	0.30
Citrus	0.20
Peach	0.25
Grapes	0.20
Mango	0.16
Coconut	0.24
Banana	O.12
Oil Palm	0.20
Avocado	0.15
Walnut	0.30
Pomegranate	0.11
Passion fruit	0.08

Apple

The deficiency is similar to N deficiency. The leaves become pale yellow, but the younger leaves get affected earlier than the older ones. Under severe deficiency conditions, the terminal buds develop the symptoms early which gives the tree a pale and more starved appearance than if the trees were suffering from nitrogen deficiency.

Banana

Sulphur deficiency appears on the youngest leaves, at first as a general chlorosis declining as the leaf ages; later, as the deficiency becomes acute, the new leaves remain chlorotic and are progressively reduced until the midrib alone remains and the meristem breaks down.

Citrus

Sulphur deficiency in citrus is similar to N deficiency. To start with, much yellower new growth is noticed in young leaves but the older leaves generally remain green. Some dieback twig occurs as the deficiency progresses. Immature fruits are lighter green and mature fruits are of the lighter orange colour. Occasionally, some fruits are mishappen and flattened, others have very thick peel. The interior of the fruit remains pulpy and juiceless with some of the juice sacs gelatinised as seen in granulation.

Cocoa

Chlorosis of the entire leaf may occur. Necrotic streaks may appear on leaf edges and leaves may fall off. Growth is stunted.

Coconut

Sulphur deficient seedlings become yellow much later than N deficient palms. The mature leaves progressively lose their green colour from the margins, so that the green seemed to form a haze on a yellow background. The rachis of the leaf is weak and pliant. Sometimes, the deficient palm leaves turn yellow and orange becoming necrotic and grey at the tips. On the older palms, the total number of living leaves is strongly reduced and the leaves are of yellowish-green colour. Sulphur deficiency palms have a marked tendency to retain their dead leaves. There is premature bending of the leaves above the normal abscission layer and consequently, a considerable number of dead leaves are hanging along the stem. Nut production is affected, the nuts being small, producing poor quality copra with low oil content.

Cherry

Due to S deficiency, the base and middle ribs of the leaves turn yellow. The discolouration then spreads along the veins to the tips.

Cashew

The deficiency has been characterised by pale green to greenish yellow discolouration of younger leaves which later turn to uniform yellow. Small necrotic spots appear on the affected leaves which coalesce to become larger necrotic areas. Shedding of affected leaves and growth retardation also occurs in the deficient plants.

Guava

The deficiency is manifested by the main veins of the younger leaves which turn purple.

Mango

Leaves show a similar appearance to P deficiency. Tree growth is retarded and defoliation begins. Mature leaves develop very deep green colour and marginal necrosis, and later fall from the tree. Sulphur deficiency scorch develops along the sides of the leaf while in phosphorus deficiency, the scorch develops from the leaf tip.

Oil palm

Small pale-green to almost white fronds show some interveinal streaking while older leaves develop necrotic spots followed by terminal necrosis.

Peach

The S deficiency in peach has been characterised by the inhibition of terminal growth and young leaves turning pale green, tips of which turn scorched and withered.

Pineapple

The S deficiency in pineapple has been characterised by the blistering on older leaves; leaves gradually become lighter green, yellow mottling with necrotic spots; plant tends to have a reddish tint.

Pomegranate

Heavy defoliation occurs at early stages. The younger leaves become chlorotic yellow. Necrotic spots develop in the margins of the chlorotic leaves.

Strawberries

Sulphur deficiency symptoms in strawberries resemble those indicating nitrogen deficiency except that, one leaflet is usually smaller than the other two. Margins of older leaves develop brownish-black serration tips.

4.2.6.6 Remedial measures

The sulphur requirement of the crops are met not only with fertilizer sulphur but also by atmospheric pollution, irrigation water, farmyard manure, crop residues and sulphur containing pesticides. However, the sulphur contribution to the crop productivity other than the sulphate fertilizer is inadequate. Fertilizer sources of sulphur contribute to a greater extent of S need of the crops.

Sulphur fertilizers may contain sulphur either as the sulphate or other soluble forms, or as elemental sulphur. Sulphur in the sulphate form has the advantage of being immediately available to the plant. However, it is also often readily leached from the root area. Elemental sulphur takes several months to be available for the plant. However, elemental sulphur has the advantage of providing a long lasting supply of sulphur to the crop. The rates generally applied are in the range of 20 to 50 kg S/ha. Among the sources of sulphur, single superphosphate is the most commonly used fertilizers. Gypsum is another important source. The sulphates of micronutrients also contribute sulphur to some extent. However, selecting the choice of sulphur containing fertilizers depends on the soil, crop, time of application, local availability and cost. Leaves absorb sulphur dioxide from the atmosphere in coastal areas as well as in industrial areas and incorporated in the plant metabolism.

4.2.7 Zinc

The use of essential plant nutrients through fertilizer is very important of get higher crop productivity. The use of high analysis fertilizers free from micronutrient coupled with intensive cropping has resulted in widespread micronutrient deficiency. Among different micronutrients, zinc deficiencies are more widespread in the field and fruit crops in different parts of India. Deficiency of any element produce characteristic effects on organ of plants and foliage characters like colour, size and shape of leaves, stem characters, like colour, fibre content, thickness, blossoming characters like size and amount and fruit characters like size, colour, hardiness and flavour. Zinc is also no exception for its deficiency symptoms. Zinc deficiency has been reported to occur about 47 per cent of the cultivatable area of our country.

4.2.7.1. Zinc Status of soil

The total Zn in soils varies from 10-300ppm with an average of around 80 ppm, occurring in number of minerals viz. Augite, hornblende and biotite. The ionic radius of Zn^{2+} is very similar to that of Fe^{2+} and Mg^{2+}. The level of Zn in soil solution is very low and ranges between 3 x 10^{-8} and 3 x 10^{-6} M. Zinc solubility is low on soils of high pH and particularly when $CaCO_3$ is present. Iron forms complex with organic matter and both soluble and in soluble Zn are formed. On an average, 60% of the soluble Zn in soil occurs is soluble Zn organic complexes. Soils originating from basic igneous rocks are high in Zn.

4.2.7.2. Factors effecting Zn availability in soils

A. Soil factors

1. Zinc solubility is highly pH dependent and decreases 100 – fold for each

unit increase in pH. The availability of soil Zn to plant decreases with increased soil pH. Its availability is high in soils with pH below 6. In alkaline soils, the availability is low and hence calcareous soils are frequently Zn deficient due to the alkaline reaction.

2. Organic matter decomposition gives rise to certain chelating agent which contribute to Zn availability to plants. Low organic matter soils are prove to Zn deficiency.
3. Coarse soil fractions contain and retain low Zn. Therefore, light textured soils are generally Zn deficient.
4. High P availability in soils induces Zn deficiency. The antagonistic effect of high P on Zn availability is aggravated in calcareous soils.
5. The highly mobile $ZnSO_4$ complex is an important species in soils and contribute significantly to total Zn in solution. Solubility and mobility of Zn^{2+} in soils are believed to be increased by the presence of SO_4^{2-} and subsequent formation of this complex.
6. Liberal application of N fertilizer can stimulate plant growth and increase Zinc requirement beyond the available supply.
7. The continuous flooding of the soil may induce Zn deficiency due to the deleterious effects of H_2S on plants under anaerobic conditions.
8. Zinc deficiencies are common during cool, and wet seasons, namely poor light as well as cool temperature and excessive moisture aggravate Zn deficiency. Increase in soil temperature have been shown to increase the availability of Zn to crops.
9. Generally, zinc is relatively immobile in soils. Most of the zinc was retained in the surface inch of soil and only little if any down ward movement occurred. Hence, any removal of top soil, either by erosion or in the process of land shapping for gravity irrigation increases Zn deficiency.
10. Species and varieties of plant may vary markedly in their susceptibility to Zn deficiency. Genotype differences in sensitivity to Zn deficiency are known to occur in several other crops.

4.2.7.3. Zinc in plants

Plants absorb Zinc mainly as Zn^{2+} ions. Zinc content ranges between 20 and 100ppm are considered as normal and of 15 – 20 ppm as critical content. In plant roots, 90 per cent of total Zn may occur on exchange sites or adsorbed to surfaces of the cell walls in the cortex. Zn uptake was considerably reduced by low temperature and metabolic inhibitors. High P content in plants inhibit the acropetal translocation of Zn^{2+}, Zinc content in the leaves are affected most, while the uptake of Zn^{2+} ions by the root is less affected. Zinc is precipated as phosphate near the vessels when phosphorus contents are high. Hence, P: Zn ratio of the plant has a major influence of both zinc status and the absolute Zinc content of the plant. Optimal values for this ratio vary from 50 to 200 depending upon the variety. Plants deficient in zinc generally has an abnormally high P content. Zinc uptake can also be inhibited by high Iron and copper content of the plant. The mobility of Fe within plant is poor, but better than that of iron, boron and molybdenum. Zinc concentration always decrease acropetally from the root, through the stem to the leaves.

4.2.7.4. Role of zinc in plant metabolism

Zinc participates in enzyme substrate binding. The number of Zinc – metallo-enzymes in plants are many, viz., carbonic unhydrase, alcohol dehydrogenase, glutamic dehydrogenase, D-glyceraldehyde-3- phosphate dehydrogenase, D-lactic dehydrogenase, aldolase, proteinases and peptidases. Among these, the most important enzyme is carboric unhydrase which catalyse the reaction.

$$H_2O + CO_2 \xrightleftharpoons{\text{carbonic unhydrase}} H^+ + HCO_3^-$$

It is localized in chloroplast. This enzyme plays an important role in photosynthesis. It also protects protein from denaturation. Enzymes involved in protein synthesis are also activated by Zn. Zinc is directly implicated in protein metabolissm. There is a strong linkage between zinc and protein content of a plant. Zinc plays an important part in nucleic acid synthesis. Zinc plays an important role in auxin metabolissm of the plant. The synthesis of tryptophane out of Indole and serine is known to be catalyzed by zinc. Therefore, zinc influences the auxin content of plant through tryptophane, the precursor of Indole Acetic acid (IAA). Zinc as a metal component of superoxide dismutase (SOD) and as an inhibitor of NADPH – Oxidase protects not only IAA but also the cell membrane.

Zinc accelerates reduction processes through its influence on the transfer of hydrogen atom from reducing to the oxidizing agent in dehydrogenation reactions. Plant lacking zinc rarely contain starch grain and chlorophyll content is abnormally low and hence leaves are chlorotic. Zinc is believed to be involved in chlorophyll synthesis through its influence on protein, carbohydrate and energy metabolism. Zinc is involved in the synthesis of vitamin B and C.

4.2.7.5. Diagnosis of deficiencies

a) Visual symptoms: Deficiency of zinc in plants produces characteristic symptoms. These can be used as a guide to diagnose the zinc deficiency.

b) Plant analysis also helps to isolate deficient crops and soils.

c) Soil analysis will indicate available Zn in soil and helps to assess the deficient nature of the soil.

a) Visual symptoms

Zinc deficiency may depress plant yields by as much as 50 per cent without producing any symptoms. Where visual symptoms develop, most often occur on leaves. At times, these can be seen on other plant parts also. Not only the overall development of plant is affected by Zn deficiency, but even root growth is retarded. Severity of symptoms can be used as an index to the degree of deficiency.

Crops suffering from Zn deficiency often show chlorosis in the interveinal areas of the leaf. These areas are pale green, yellow or even white. In fruit trees, leaf development is adversely affected. Unevenly distributed clusters or rosettes of small stift leaves are formed at the ends of the young shoots. Frequently, the shoots die off and leaves fall prematurely. The bark of the tree are rough and brittle. Symptoms due to auxin deficiency are characteristics of zinc deficiency. These include 'Little Leaf' accompanied by more or less distinct leaf deformation, stunted growth and resetting owing to short internodes. Deformation and premature dropping of fruits are also typical symptoms. Since auxin decomposes rapidly at high light intensities, zinc deficiency symptoms are particularly striking in plants exposed to intense light. The development of "rosette disease" is encouraged by clear sunny weather and zinc deficiency symptoms are therefore much more pronounced or may even develop only on the sunny side of a tree. This is also why zic deficiency is more common in dry regions with strong isolation. Zinc deficiency symptoms of the some of the important fruit crops are furnished below.

Banana

Mild deficiency resembles the effect of N shortage; severe deficiency produces punchy foliage, stunted growth, leaf deformity and increase in the length/breadth ratio of the leaf. The leaves become either entirely white or white towards the base with the rest of the blade showing chlorotic transverse strips alternating with greener bands. The chlorotic parts showed numerous small, oblong, light patches which ultimately rotted. In severe deficiency, the fruit development is slow and the fingers are twisted shorter, thinner and light green in colour than those of normal plants. Alkalinity and high soil phosphate probably contribute and the symptoms are more acute on the sides of plant facing the sun.

Citrus

Zinc deficiency is variously called "Frenching", "mottle leaf" and "folio callosis". The trees will look like bushy appearance. Leaves develop partial chlorosis between the veins, while parts adjoining midrib and lateral veins remain green. Degree of chlorosis increases with severity of deficiency. Leaves become small, pointed and narrow in acute cases, but of normal size in mild cases. Young growth is affected on occasional weak twigs. Deficiencies are intensified by damage to the roots caused by implements, fungal and virus infections may also aggravate them by attacking the roots and blocking the conducting tissue.

Grape

Symptoms are found in vineyards on sandy soil and sometimes application of excess phosphorus leads to zinc deficiency and more pronounced in hot, high sunlight areas than in cooler coastal regions. Symptoms begin to appear when the secondary or auxiliary shoot growth begins. The general symptoms are chlorotic pattern, reduced leaf size, formation of wide angle between the two basal lobes of the leaf blade when the petiole remaining attached, the primary veins tending to draw together, resulting in a fan-like appearance with small underdeveloped.

Pineapple

Small yellowish spots first appear on the upper surface of the adult leaves, which enlarge and become brown at the centre, coalesce and ultimately cause complete destruction of leaf tissue.

Avocado

Zinc deficient plants show pale coloured small leaves, multiple buds and broom like appearance of terminals. Plants also produce interveinal yellowing of leaves and formation of rosette. The mature leaves often develop small necrotic spots on the blades.

Apple

Zinc deficiency of apple is commonly called "rosette" because of the characteristic symptoms. Rosette is characterized by a dense cluster of small narrow leaves, that terminate a branch, which is usually leafless or with space below the rosette. Branch terminals are often affected first and then the laterals where the leaves are transitional in colour and shape between the normal leaves and the almost sessile, narrow small leaves of a rosette. Rosette first appear on the spring growth. If resetting occurs late in the season, the symptoms are less pronounced, but such shoot terminals may die in the following spring after a week of starting a weak growth, leaving behind a number of short lateral spurs. These shoots are susceptible to winter injury than normal shoots.

Peach

Zinc deficient plants show mottling of younger leaves, which ultimately become small, narrow, pointed and often had a wavy margin. Defoliation of leaves also occurr from the upper mid portion of shoots and the roots show gummy deposits. The little leaf and rosette condition of trees are the important characteristic symptoms of zinc deficiency. Deficient trees produce misshapen pointed small fruits with flat cheeks.

Pear

Zinc deficiency of pear is more or less similar to apple. Prominent deficiency symptoms are little leaf, rosette and leaf mottle.

Cherry

The zinc deficiency symptoms in cherry is characterized as little leaf. The leaves, which are usually misshappened, become small, narrow and chlorotic. Chlorosis appears first on the leaf margins which gradually progress towards the midrib and the large interveined areas, while the midrib and lateral veins still remain green. One of the most distinctive symptoms of Zn deficiency is the appearance of small, narrow leaves in short rosette in spring. The leaves may also show characteristic mottling with yellow streaks and splashes between the veins. The size of the fruit is reduced.

Strawberry

Zinc deficient plants exhibit pale green to yellow color of the leaves. Only the larger veins and a narrow strip along the margins remain green. The leaves grow slowly while the leaf tend to be some what narrow and concave. The deficiency is more pronounced on younger leaves.

Pecan

Zinc deficiency is called rosette in pecan, which is common in places with coarse textured sandy soils. The most striking symptoms seen on the severely affected trees are the occurrence of rosette of very small wrinkled leaves and die back of limb ends. Often the trees are stunted or even killed. In mild cases, the chlorotic condition between veins increase, bronzing of trees as the season advanced, and small size of fruit might be the most important symptom. Severely deficient trees produce little or no crop and nuts are of much smaller size than the normal ones.

Walnut

Zinc deficiency, giving rise to "Yellows" or 'little leaf' in walnut is common in areas where soils are coarse and sandy. Walnut trees affected by zinc deficiency showed small leaves with interveinal chlorosis. Sometimes die back of branches may also occur. Moderately zinc deficient trees may have normal leaves in early spring, which may turn yellow or curl in late spring or early summer.

Almond

Zinc deficiency symptom is called "little leaf" in almond. Almond trees affected by Zn deficiency showed small leaves, chlorosis between veins, increasing towards leaf tip and resetting in several cases, but most of the orchards affected are characterized commonly by pale leaves and poor crops.

Mango

Zinc deficiency produces stunted growth of roots, shoots and leaves and the leaves turn pale yellow but to a lesser extent than manganese or other elements. The area along midrib and main laterals, however remain green. Leaves become very small, thin, brittle, narrow and pointed and the petiole makes acute angle with stem. Yellowing necrotic patches develop on old leaves with drying of leaves. Subsequently, necrotic patches turn grey and brown and cover the entire leaf surface. In some cases, dead necrotic tissues fall off leaving holes.

Guava

Deficiency of zinc causes small leaf and chlorosis, Interveinal chlorosis is observed with a reduction in leaf size, poor growth, die-back, less flowering and drying and cracking of fruits which sets.

Plum

Due to zinc deficiency, small, pointed and often boat shaped leaves are produced. When the zinc deficiency is severe, terminal growth is stunted and leaves are produced in rosettes. Affected foliage has a characteristic interveinal molting where the midrib and lateral veins stand out green against a lighten background. Some varieties of plum show stripped appearing foliage due to small islets of chlorotic areas scattered on the leaf.

Black currants

Zinc deficiency produces very small, scarcely unfolded leaves that grow together in dense clusters (rosetting) owing to inhibited internodal growth.

Sapota

Zinc deficiency results in reduced length of inter nodes, small and erect leaves, defoliation on terminals and poor flowering.

Coconut

Zinc deficiency results in dwarfed or deformed young leaves and the classic 'Little leaf' condition.

Oil palm

Zinc deficiency is not common in oil palm but may be induced under high soil P. Deficiency results in small, narrow white streaks on lower and mid-crown fronds. A different condition that produces blotchy leaf symptoms has also been identified as Zn deficiency.

The leaf is the site of many physiological activities of plants. Besides being the site for production of carbohydrates in photosynthesis, the leaves play a vital role in metabolism of most of the plant constituents. Nutrient deficiency and toxicity are usually expressed by leaves. Nutrients absorbed in excess are normally accumulated in leaves and thus leaf forms an ideal plant part for nutrient diagnosis. The critical concentration of zinc also varies with the fruit crops.

Table 12: The critical content of zinc in fruit crops.

Crops	Zinc content (ppm)
Mango	12.0
Grape	13.0
Citrus sp.	15.0
Avocado	10-20
Apple	15.0
Peach	12.7
Pear	9.9
Pecan	40.0
Almond	36.0
Papaya	17.8
Oil palm	15.0
Walnut	15.0
Banana	13.4

4.2.7.6 Amelioration Techniques

In view of the widespread deficiency of Zn, its amelioration has become a serious national concern. Also due to an unprecedented rise in the amelioration costs, selection of cheaper sources and innovation of more efficient methods of application has attained top priority Hepta hydrate Zinc sulphate ($ZnSO_4$.$7H_2O$) is the commonest of the Zn carrier which has been tested to overcome Zn deficiency. Zn could be provided to deficient crops by following methods.

- Soil application – broadcost or banded
- Foliar application as spray
- Dusting seeds with Zn powder

Among these methods of Zn application, soil application and foliar spray are frequently used. In citrus, Zn deficiency can be corrected by spraying 0.6% $ZnSO_{4.}$ In banana, 0.5% $ZnSO_4$ is found effective in reducing the zinc deficiency. In general, in fruit crops, foliar sprays of 0.6% $ZnSO_4$ + 0.3% lime one at one week before bloom and another at after bloom gives good results. Spraying 0.8% Zn SO_4 improves the quality of fruits in addition to the correcting deficiency in many fruit crops. Cholortic leaf in both pineapple as well as in guava can be successfully restored to health by spraying 0.4% solution of zinc sulphate. In the case of apple, trunk injection of dry Zn SO_4 has been found to be effective in controlling Zn deficiency. However, the method is time consuming and causes injury and hence not recommended. In peach, Zn deficiency can be controlled by spraying 0.4% $ZnSO_4$. Zinc sulphate spray of 0.2%, applied during early growing season is found effective in pear.

4.2.8 Copper

Copper is an important element in fruit production and its deficiencies have increased since the introduction of high yielding fruit trees. It's essentiality for plant growth was established by Sommer *et al.* in 1931. Copper deficiency in soils has been observed to be maximum in Kerala (31 percent) and minimum in Punjab (less than 0.1 percent) with an all India average of 5.0 percent. Even though copper is required in very small amount, it is very much essential for the growth of plants either as constituent or required for functioning of several important enzymes which regulate many metabolic pathway. It's deficiency causes 20 to 100 per cent yield loss depending upon the severity and persistence of the deficiency.

4.2.8.1 Copper in soil

Copper occurs in the soil almost exclusively in divalent form. Copper is usually present in the crystal lattices of primary and secondary minerals. Copper is strongly bound to organic matter and Cu organic complexes play an important role in regulating Cu mobility and availability in the soil. Cu availability generally decreases as the pH of the soil increases due to liming. Copper added to the soil is restricted to the upper soil horizons and therefore, the Cu content of many soils decreases down the profile. Copper deficiencies are likely to occur in humus rich soils due to strong binding of Cu^{2+}. Copper availability in organic soils depend not only on the concentration in soil solution but also on the form in which the Cu occurrs. Copper complexes in the soil solution of molecular weight <1000 are much more available to plants than Cu completes with a molecular weight is in excess of 5000. Copper content in soils range between 10 and 200 ppm with an average of 50 ppm. Deficiencies are likely to occur when the total Cu content is less than 6 ppm in mineral soils and below 30 ppm in organic soils.

4.2.8.2 Factors affecting copper availability in soils

1. Copper level varies with the changes of soil texture, it's content increasing with an increase in the fineness of soil texture. Sandy soils are naturally low in Cu and are more frequently Cu deficient due to excessive leaching.

2. The concentration of Cu in soil solution decreases with increasing pH and its supply to plant is reduced because of decreased solubility and increased adsorption. Its availability is reduced above pH 7.0 and it is most readily available below pH 6.0. Low availability of Cu at low pH may be due to lock up with aluminum silicates and phosphate inos and also due to the excessive uptake of other nutrient ions which hinder the uptake of Cu adsorption.

3. Copper is strongly bound to organic matter. Hence, Cu deficiencies are most common in organic soils. The deficiencies are accentuated if poor drainage condition also prevails. In arid regions, Cu deficiency occurs in soils consisting of little organic matter.

4. Application of N-P-K fertilizers can induce Cu deficiencies. Copper deficiency also may be induced due to increased Al^{3+} by the use of acid forming N fertilizers in acid soils. High concentration of Zn, Fe and P in soil solution also depress Cu adsorption by plant roots and may aggravate Cu deficiency.

5. Copper availability in soil is generally associated with the total Cu content. Soil analysis of Cu indicates the sufficiency or deficiency of Cu in soils. Mineral soils that contain less than 6 ppm or organic soils which contain below 30 ppm total Cu are looked upon with suspicion for deficiency.

6. Crops vary in sensitivity to express Cu deficiency. The genotypic differences in Cu nutrition among crops are related to many factors. Severe Cu deficiency is observed in crops planted in high C:N ratio.
7. Incorporation of unharvested brassica root crop residues into the soil aggravates Cu deficiencies in the following crops which is probably related to the larger amounts of S released during decomposition of these residues.

4.2.8.3 Copper in plants

Plant absorb copper as free Cu^{2+} ions and probably also in the form of chelates through the roots. Copper content in plants generally varies between 5 and 20 ppm. Copper uptake appears to be metabolically mediated process. The uptake of Cu^{2+} ions by plants is very low compared to that of boron, iron, manganese and zinc. Copper content is the highest when the plants are young which decrease towards maturity. It behaves like a mobile element in Cu sufficient plant and similar to an immobile one in Cu deficient plants. So in plants well supplied with Cu, the shoot tips and young leaves will have relatively higher concentration of Cu than old leaves. In comparison, young leaves of Cu deficient plant will usually contain lower Cu than their old leaves. Roots contain higher Cu content than the other plant tissues. Copper has a strong affinity for the N atom of amino acid and act as a Cu carrier in the transport of amino acids through xylem. The uptake of copper at optimum yield levels varies between 30 and 100g//ha and at high yields vary from 120 and 150 g/ha.

4.2.8.4 Functions of copper in plants

Copper is either a constituent of or associated with many compounds which are of functional importance in the metabolim of higher plants. The physiological effects of Cu^{2+} ions are closely linked with their ability to change valency ($Cu^{+} \rightleftharpoons Cu^{2+}$)and its capacity to form stable complex compounds. Above all, copper has a strong affinity for certain protein structures in which Cu^{2+} ions are bonded tightly to the protein structure. The leaves that are rich in protein generally have the highest copper content. Nearly 70 percent of the copper in plants is situated in the grana of the chloroplast as a constituent of plastocyanin (blue protein), the terminal link acting as an electron donar in the chloroplast – electron transfer chain in photosynthesis. The plastocyanin and copper content of chloroplast is up to 50% lower than normal in copper deficient plants. The Cu^{2+} ions are believed to act as chloroplast stabilizer, protecting the chlorophyll – protein lipoid complex and chlorophyll from premature degeneration and thus it delays senescence. Copper is also implicated in the second stage of nitrate reduction i.e., in the reduction of nitrite.

Copper is a constituent or activator of enzymes involved in respiratory metabolism and electron transport, including lactase, polyphenol oxidase, tyrosinase and ascorbic acid oxidases. Copper containing enzymes are implicated in oxidation – reduction processes in which O_2 as an electron acceptor is reduced to H_2O_2 and H_2O. The enzyme polyphenol oxidase and tyrosinase are involved in the secondary metabolism of plants. They are not only maintaining the delicate balance of phytohormone, the colour of the flower but also in defense responses to fungal and viral infection. Copper also enhances the cytochrome activity. The importance of copper for respiratory metabolism is also shown by the fact that the germination capacity of copper-deficient seeds is very poor. Copper is also implicated in the biosynthesis of lignin, and thus in the stabilization and lignification of cell walls. Lignification can be inhibited or completely absent in Cu deficient plants leading to bending and distortion of leaves and stem. Xylem vessels may collapse due to lack of structural thickening, leading to restricted water movement and hence wilting. Pollen sterility has been found in a number of plant species grown under copper deficiency. Pollen sterility can be linked with a tapetal abnormality in anther during pollen development.

4.2.8.5 Diagnosis of visual symptoms

Deficiencies occur when a soil is unable to supply adequate concentration of copper for uptake by the growing plant. Soils with very low concentration of available copper will cause severe deficiency in fruit crops manifested by visible symptoms. Yield loss in such cases can range from 20 per cent to 100 per cent with quality also being affected However, where the capacity of the soil to supply copper is marginally inadequate 'Latent' or subclinical deficiencies can occur, where yields are reduced upto 20 per cent without the appearance of obvious visible symptoms. This later type of deficiency is difficult to detect and therefore of considerable economic importance. The combination of soil factors responsible for the deficient supplies of plant – available copper are relatively limited and once the deficient soils have been identified, their treatment is easy and effective. The susceptibility of crops to copper deficiency varies between major plant groups and even between cultivars. Among the fruit crops, citrus crops are highly sensitive to Cu deficiency; whereas apple, peaches and pears are moderately sensitive to Cu deficiency. Copper is taken up by plants in the form of Cu^{2+} ions by active absorption. Copper deficiency is always found to develop first in younger leaves.

Copper deficiency can be identified by the recognition of visual symptoms or by soil or plant analysis. The use of visual symptom is the quickest and cheapest approach. But one should not confuse with the symptoms of those of other nutrient stresses or damage, the subclinical (Hidden hunger) may not be easily detected. It may sometimes be too late to correct the deficiency in an existing crop owing to the symptoms becoming apparent only at a fairly advanced stage of growth. Plant analysis is more reliable than the diagnosis of symptoms and would be expected to be better than soil test because it indicates the copper status of the crop plant itself under a given set of soil and environmental conditions. However, timeliness can sometimes also be a problem with this technique because specific tissues or whole shoots have to be sampled at a stipulated stage of growth and it may be difficult to collect all the sample at the appropriate time. Delays in analysis may prevent corrective measures being taken in time to correct any deficiencies detected.

Effective use of plant analysis depends on an understanding of the variations in copper concentrations within different crops with stages of growth and their correction with yield. Consequently, standardized sampling, absence of contamination and the use of locally – appropriate critical values for interpretation of analytical data are necessary. Copper content below which deficiencies normally occur varies among crops.

Table 13: Critical concentration of copper in plants

Crops	Cu content (ppm) less than
Coconut	3.0
Oil palm	3.0
Citrus	3.6
Grapes	4.0
Walnut	4.0
Apple	7.0
Pear	5.0
Peach	6.0
Apricot	5.0
Cherry	5.0
Almond	5.0

Generally, symptoms appear when the leaf copper level is below 4 ppm. Since the initial symptoms and those associated with mild deficiencies are not so pronounced as those of Mg, B or Fe deficiency, copper deficiency is often difficult to identify.

Apple

Terminal branches die back. Terminal leaves develop necrosis followed by death of shoot tips. The next year growth is resumed by buds below the point of death and die back is repeated. This annual occurrance causes affected trees to have a bushy, stunted appearance.

Citrus

Abnormally large, dark green leaves give early indications of deficiency. Twigs die back, and multiple buds form near the juncture with live wood. The most reliable symptoms for identification of copper deficiencies are gum pockets under the green bark of young wood and brownish excrescence on fruits, twigs, and leaves, New shoots may produce narrow and elongated leaves. In acute deficiency, fruits develop brown lesions, the rigid peel may split and fruit drops.

Mango

Shoots produced on long drooping, S-shaped branches of previous growth are weak, lose foliage and die back.

Pineapple

Young leaves narrow with curled edges. Initially, leaf tip starts drooping and later become necrotic. Lower surface of the leaves become waxy and lack bloom.

Apricot

Terminal branches die back at tips preceded by cessation of terminal growth. Rosette formation and multiple bud growth on terminals occur; leaves turn pale green to bright yellow in interveinal areas.

Pear

Terminal tips of current season's growth die back progressively downward. Leaves die at tips and margins towards centre. Orange brown striations occur parallel to margins. Trees bear no fruiting.

Coconut

Severe bending of rachis of the youngest leaves accompanied by yellowing and drying of leaf tip which appears rimmed with brown and yellow, while the middle portions remain normal green. As the symptoms develop, new leaves are deformed and are abnormally short giving the coconut a rusty sagging appearance.

Banana

The deficiency causes development of ribbon like roots and is characterized by internal chlorosis, leaf tip chlorosis, rosetting, increased lamina and withering and drying of leaves, poor growth of stem, detopped appearance, development of strong water shoots. Die back and defoliation with or without eruptions occur. The fruits are with irregular brown spots with splitting of the skin.

Plum

The leaves are dark green at first. Later the younger leaves turn yellowish green to yellow between the nerves, large parts of them then taking on a reticulate appearance. The branches are later denuded.

Peach

The foliage of the tree is initially, unusually dark green in older leaves, but later the leaves turn greenish-yellow between the fine veins until finally a green reticulate pattern is left on a whitish-green back ground. The young, long and narrow leaves are much smaller than normal, deformed wavy and have irregular margins. They soon drop, leaving the branches bare as in other fruit trees.

Strawberry

The leaves develop intercostal chlorosis and lighter blotches along the leaf margins and near the middle rib. The leaves are smaller than usual, collapse and wilt.

Papaya

Reduced leaf size is observed particularly in young and middle leaves. The young leaves show diffused general paling. The epical half of the lobes developed minute brown spots in the interveinal area. In acute cases, some of the affected leaves lost turgor dried and had a wilted appearance.

Almond

Stunted tree growth. Gum exudates from the trunk during late winter and spring appear. Deficient trees have rough bark on the trunk and older branches which later turn black. On some leaves, there is interveinal yellow mottling, more prevalent towards the apex and near the margins but sometimes evenly distributed over the leaf. On others, there is yellowing in a more or less district, longitudinal band close to and paralleling the midrib. Marginal and apical scorch is sometimes associated with those symptoms.

Walnut

Copper deficiency symptoms appear in mid-summer. Scorch develops on leaves near the tips of the shoots occur and small dark brown spots appear on the shoot near the tip. Terminal dieback subsequently occurs in late summer. Kernels are often badly shrivelled.

Pomegranate

Leaves are dark green with waxy appearance, with necrotic apices and plants have stunted growth.

Avocado

Older leaves have a dull appearance, with the veins becoming reddish-brown colour which gradually spreads into the leaf blades; there is multiple bud formation at tips of twigs. Abortive new leaves form, which almost immediately begin drying up and dying back until the entire twig dies.

4.2.8.6 Correction of deficiencies

Copper deficiencies in crops can be corrected either by increasing the concentration of available copper in the soil through the use of copper fertilizers, or by supplying the element directly to the plant foliage. The most frequently used method of correcting copper deficiency is to apply copper sulphate ($CuSO_4 . 5H_2O$) to the soil. The copper sulphate can either be broadcast evenly over the surface and incorporated into the soil by cultivation, or banded in the vicinity of the seed, either before or at the lime of sowing. Copper compounds are frequently mixed with macronutrient fertilizers to faciltate application.

Wide variations exist in the amount of copper applied to soils for the correction of deficiencies. Differences in the placement of copper in the soil in relation to plant roots, adsorption of copper by soil constituents, reaction of copper fertilizers with soil constituents affecting plant uptake, and in the copper requirements of plants can explain some of these variations. There are variations in the recommendations. In general, application of copper to organic soils tend to be larger and more frequent than those to mineral soils. For soil application, 5.5 to 28.0 Kg /ha of copper sulphate is enough and one soil application of Cu produces a strong residual effect which may last for 2-8 years. Organic soils need still higher amount than the other soils. Organic manures rich in copper can also be applied to increase the soil available Cu. Among the manures, pig manures, was considered to be the richest source of copper.

The residual effect of soil applied Cu is due to the strong adsorption and organic complexation of copper by soil constituents and poor initial mixing within the rooting zone. In following years, improved mixing by cultivation and gradual desorption of copper ions provide an increased copper supply capacity.

Foliar application of copper is also practiced and is usually used to rectify the visual symptoms of Cu deficiencies. Copper can be applied in sprays mixed with other agrochemicals thus dispensing with an extra spraying operations solely to apply copper. Foliar application of 0.1% copper sulphate is recommended. This spray solution can be mixed with 0.05% lime in order to avoid scorching. Copper sulphate solution of more than 0.2% causes phytotoxicity due to its high solubility. Hence, copper oxychloride and chelates can safely be used as foliar sprays. The Cu requirement of foliar spray per hectare ranges between100g and 500g depending upon the crop and the intensity of the symptoms.

Foliar application ensures that each crop receives adequate supply of copper which may be less certain with soil treatment owing to sorption of the metal. There is also a beneficial fungicidal effect to be gained from the presence of copper compounds on foliage. The timing of foliar application is also very important. Foliar spray at the formation of new flesh ensures maximum and considered as the best time for foliar spray. Two to three spray can be done depending upon the severity of deficiency. Besides spraying copper sulphate and copper oxychlorides as measures to rectify Cu deficiency, spraying Bordeaux mixture to control fungal diseases is also effective in controlling copper deficiency.

4.2.9 Iron

Iron was the first micronutrient to be discovered as an essential element for plant growth by Gris in 1844. Plants absorb Fe^{2+} ions by young lateral roots by active processes. The absorption of Fe decreases as the age of the crop advanced. Iron is less mobile in plant and it is translocated in the form of soluble ferrodoxins and citrate complexes. Iron content in plant tissues range between 50 and 500 ppm. Among the crop plants, fruit crops and ornamental are susceptible to Fe deficiency. Citrus, grapevine, peach, pear, plum and apple are the some of the fruit crops and ixora and palms are the some of the ornamental crops susceptible to iron deficiency.

4.2.9.1 Iron in soils

Iron constitute about 5.0% of the earth's crust and is invariably present in all soils. Soil Fe usually occurs in the crystal lattices of numerous minerals viz., olivine, biotite etc,. The total iron content in soils vary from 0.02 to 10.0 percent. The content of soluble Fe in soil is extremely low in comparison with the total Fe content. Soluble inorganic forms include Fe^{3+}, $Fe(OH)^{+}$, $FeOH^{2+}$ and Fe^{2+}. In well drained oxidized soils, the solution Fe^{2+} concentration is less than that of Fe^{3+} . At higher pH levels, Fe^{3+} activity in solution decreases 1000 fold for each pH unit rise. In contrast, Fe^{2+} decreases 100 fold for each unit increase in pH. Acid soils are thus relatively higher in soluble inorganic Fe than calcareous soils where Fe availability is extremely low. When soils are waterlogged, a reduction of Fe^{3+} to Fe^{2+} takes place accompanied by an increase in Fe solubility. Reduction is brought out by anaerobic metabolism of bacteria. Insoluble Fe^{3+} form predominates in well drained soils and this is due to the conversion of $Fe^{2+} \rightarrow Fe^{3+}$. It has been reported that at a depth of 2 m, over 90% of the soluble Fe is present as Fe^{2+}. In the deeper soil layers which are less aerated, the fraction of Fe^{2+} of the total soluble Fe is frequently higher than in the upper horizon. The reduction potential generally falls from upper to lower horizons.

4.2.9.2 Iron in plants

Iron deficiency is estimated to be 11.2 percent of the soils in All India average and it is also predicted that further increase of Fe deficiency may occur in the years to come. The critical concentration of Fe in soil is found to be 3.7ppm (DTPA) for non-calcareous soil and 6.3 ppm (DTPA) for calcareous soil. The Fe deficiency is exhibited in plants when the Fe content is below 50 ppm. It has been estimated that only 10 to 20 percent of the total Fe content of the plants are physiologically active. It was revealed that as much as 80 percent of the total Fe in plants is present in chloroplast and this is because of Fe deficiency results in chlorosis. Fe deficiency not only causes reduction in fruit yield (upto 40 percent), but also affects the fruit quality.

Generally, iron may be absorbed by plant roots as Fe^{2+} or as Fe chelates and Fe^{3+} is only of minor importance because of low solubility. The availability of inorganic Fe to plant roots there fore appear to be dependent on the ability of roots to lower the pH and to reduce Fe^{3+} to Fe^{2+} in the rhizosphere. Fe – cheletes are more soluble but the uptake is very low. For utilization of chelates, separation of Fe and organic legand (chelate splitting) has to take place at the root surface. Crop species differ in ability to utilize sparingly soluble inorganic Fe and Fe cheletes for iron nutrition. Fe efficient crop species are able to lower the pH of the nutrient medium and to increase the reducing capacity of the root surface under iron stress condition where by Fe availability and absorption by the roots increases tremendously. Fe is translocated in the xylem as ferric citrate.

4.2.9.3 Functions of Iron in plants

Iron plays an important role in many physiological and biochemical processes of crops. Iron is an essential component of many enzymes viz. catalase, peroxidase and cytochrome oxidase as prosthetic groups. Iron forms part of the porphyrin ring which is the structural component of cytochromes. Cytochromes are constituents of the redox systems in chloroplast and mitochondria. Iron is essential for the synthesis of leghaemoglobin in root nodules of legumes. These leghaemoglobins are required for N-fixation. Iron is also important for the conversion of nitrate to ammonia. It forms the component of ferredoxin which is needed for NADPH production in photosynthesis. Iron is also important for the activity of superoxide dismutase, succinic acid dehydrogenese, glutamate synthetase, sulphite reductase, xanthine oxidase, nitrate reductase and aconitase. Iron is indispensable for chlorophyll synthesis. Iron is need for the formation of protochlorophyllide from Mgprotoporphpyrin. Iron is also essential for the activity of coproporphyrinogen oxidase and this enzyme is essential for chlorophyll synthesis. Iron is also needed for protein synthesis. Ethylene formation is mediated by Fe. Iron is also the component of the enzyme lipoxygenases which catalyse the peroxidation of long chain polyunsaturated fatty acids. This enzyme activity is essential for fast growing tissue and also for disease resistance.

4.2.9.4 Factors conducive for Fe deficiency

1. Fe availability is very low in sandy soils
2. The availability of Fe decreases with the increase in soil pH more than 7.4. Higher the exchangeable sodium, lower will be the Fe availability.
3. Organic matter deficient soils are low in available Fe.
4. The presence of proportionately higher ions of P, Zn, Cu, and Mn promotes Fe deficiency.
5. Poor soil areation, compaction and salinity also induces Fe deficiency.
6. Removal of topsoil due to land levelling or soil erosion results in Fe deficiency.
7. Extremely high or low temperature favours Fe deficiency.
8. The infection of root nematode reduces the Fe absorption and results in Fe deficiency.
9. The irrigation water with higher bicarbonate ion also induces Fe deficiency.
10. Fe availability is very poor in calcareous soils. The calcium carbonate in the soil produces carbonic acid.

    $$CaCO_3 + CO_2 + H_2O \rightarrow Ca^{2+} + 2HCO_3^-$$

    HCO_3^- (bicarbonate ion) induces high pH and in turn affect the reduction of Fe^{3+} to Fe^{2+} before it is taken up by the plants. Thus the calcium carbonate ($CaCO_3$) induces Fe deficiency even though the soils contain Fe. Among the factors, the presence of calcium carbonate is considered to be the most important factor responsible for Fe deficiency.
11. Application of high levels of nitrate N induces Fe deficiency.
12. Infection of plants with virus also induces Fe deficiency

Table 14: The critical Fe content in various fruit crops.

Crops	Fe content (ppm)
Mango	171
Citrus	36
Grapes	35
Apple	120
Pear	100
Peach	100
Coconut	45
Avacado	40
Strawberry	50
Sapota	20
Pomegranate	21
Guava	25
Banana	35
Gooseberry	18

4.2.9.5 Diagnostic technique

Diagnosis by visual symptoms is easy, but it requires a special skill and constant touch with field problems. The deficiency of a particular element makes an ionic imbalance in the plant tissue which intern altered the physiology of the plant. The altered physiology is transformed morphologically as deficient symptoms.

The characteristic symptoms of Fe deficiency are closely linked to the disturbances of the main metabolic reactions influenced by Fe. Iron deficiency inhibited transfer of energy needed for growth processes and synthesis of chlorophyll due to lack of Fe distribution. In the event of a slight deficiency, the youngest leaves are initially yellowish green and the intercostal areas then turn lemon – yellow or an orange like colour as the deficiency becomes more severe. The nerves always remain green with sharply defined margins. The notable feature of Fe deficiencies are, in the younger leaf, the more intensive and characteristic chlorotic symptoms which can also be spread to the nerve if the deficiency is severe and persistent. In this case, leaves may became uniformly chlorotic without collapsing. The unfolding leaves of severely deficient plants are yellowish or almost completely white or only the nerve at the leaf margins are green with clearly defined contours. The older leaves still be completely normal green. In general, growth is inhibited and flowering is affected depending on the severity of chlorosis. New leaves often remain small, less fruits appear and the fruits are poorly coloured.

4.2.9.6 Visual symptoms

Citrus

Young leaves are mostly affected while the older leaves remain green. The affected leaves exhibit a network of green veins on a very light green background. In advanced stages, the veins also loose green colour. Under continued deficiency, the leaves become progressively smaller and eventually the new growth ceases. The affected leaves are shed, so that the lower portion of trees has a fair amount of good foliage while the tops may show only sparse foliage and defoliated twigs. Under acute cases, the fruits are small, hard and light in colour.

Strawberry

The young and youngest leaves are yellow to yellowish-white with green sharply defined nerves. In addition, the margins of older leaves turn brown and die. The veins of the younger developing leaves also become chlorotic if the deficiency persists.

Peach

The deficiency is characterized by an interval chlorosis which occurs in April when leaves have almost expanded and trees are making rapid vegetative growth. The new flush, which occurs in July-August also, shows similar chlorosis. Symptoms are evident as interveinal chlorosis of top leaves; the fine veinlets and main veins in the leaf blade remaining green. As the chlorosis progresses, the affected areas of the blade lose green colour, they bleach out and in extreme cases, the leaves may become ivory white; the leaves drop prematurely and eventually the denuded shoots die back.

Apple

Iron deficiency exhibits unique pattern of chlorosis of the leaves without affecting the veins. Thus, the veins form a green network on the leaf. In acute cases, the whole leaf becomes white. Later, the area along the margins of the chlorotic leaves die out, forming brown patches. Shoots and branches may die back and eventually the whole plant my die. The symptoms of chlorosis may also occur in the fruit or the fruit may be highly fleshed and low in sugars.

Banana

Interveinal chlorosis of the young leaves, a general chlorosis if deficiencies become severe, is the main symptom in banana due to Fe deficiency.

Walnut

The entire leaf is uniformly yellow with the terminal leaves affected more than the basal. Sometimes the leaves are so chlorotic that they appear white. Often basal leaves on shoot are green and only the leaves towards the terminals are chlorotic. In severe cases, the chlorotic leaves may develop scorch and abscise.

Sapota

Deficiency is noticed in young plants grown in calcareous soils, resulting in bleached appearance in terminal leaves.

Cherry

Terminal leaves become chlorotic; branches die back in severe cases.

Gooseberry

Terminal leaves are slightly chlorotic.

Pineapple

Terminal growth chlorotic, yellowish white, beginning at margins; leaves die back at tips and dark brown spots may occur in the whole plant.

Pears

The deficiency is characterized by a distinct pattern of chlorosis, developing first on the actively growing leaves on shoot tips. Major and minor veins remain green while interveinal tissues turn chlorotic to near white. Necrotic regions may develop along the margins or interior of leaves. Severely affected leaves abscise prematurely and shoot extension and diameter are reduced.

Guava

The terminal leaves show mild interveinal chlorosis. Later, it spreads to the lower leaves. In acute cases, the whole leaf becomes white.

Grapes

The loss of chlorophyll starts between the small leaf veins. The blades show the most severe fading at their margins and from there the yellowing enters the interveinal areas. The edges of severely yellowed leaves slowly dry and the dead tissue rolls upwards. Finally, the leaves dry and fall off. Lateral shoots develop with short, thin pink-coloured internodes. They have small, almost totally bleached narrow, sharp-toothed leaflets that are folded along the central veins. In acute cases, the delicate leaflets dry up, and eventually the shoots also desiccate starting at the tip.

Plum

Leaves near tips of shoots become chlorotic; detailed vein pattern may not be visible; branches die back when the deficiency become severe.

Coconut

Gradual yellowing of entire leaflets, in longitudinal strips parallel to the veins, with leaf completely yellow in its advanced stage. The rachis and leaflets become shorter with a rusty appearance.

4.2.9.7. Remedial measures

Iron deficiency can be corrected not only be supplementing the iron sources but also by managing the other factors which are conducive for iron deficiency. If deficiency is due to insufficient iron supply, different sources of iron can be applied either as soil application or foliar spray.

Among the different sources, Ferrous sulphates ($FeSO_4$. $7H_2O$) which contain 20.5 percent iron, is the cheapest source of all and is the commonly used. Iron chalets are most efficient but they are costly. Organic manures are potential sources of iron. Goat manure, Farmyard manures, cowdung and poultry manures are rich in Iron. They are known to enhance the availability of native Fe and increase the efficacy of added Fe. For soil application, ferrous sulphate at the rate of 50-250gm/ tree along with farm yardmanure can be applied depending upon the size of the canopy and the severity of the deficiency symptoms.

In problematic situations, foliar spray is cheap and effective. Foliar spray of 0.5 to 2.0 percent ferrous sulphate at the rate of 400 litres of spray fluid per ha, is recommended. A wetting agent at the rate of 1ml per litre of spray fluid improves the effectiveness of the spray. Spray can be repeated 2 to 3 times preferably during morning hours with an interval of 10-14 days depending upon the severity and persistence of symptoms. Iron can also be supplied to the plant through drip irrigation.

4.2.10. Manganese

Manganese occurs in soil as simple oxides, complex oxides, carbonate and silicate. It occurs in nature in di-tri and tetra valent forms. Divalent Mn^{2+} is adsorbed to clay minerals and organic matter and is also the most important Mn form in plant nutrition. Manganese deficiency has been estimated to 4.0 percent in All India average with a maximum of 19.0 per cent in Karnataka followed by 8.0 per cent in Tamil Nadu. The essentiality of Mn for plant growth was established by Mc Hargue in 1922. Compared to the cereals, fruit crops are more sensitive to Mn deficiency particularly citrus, apples, peaches, grapes and strawberry. The importance of manganese is felt in fruit crops and the deficiencies are widespread throughout the country.

4.2.10.1 Manganese in soils

Manganese concentration in the earth's crust average is 1000ppm and found in primary rocks viz. Ferro- magnesian materials. Total Mn in soils generally range between 20 and 3000ppm and an average of about 600ppm. Mn exists as solution Mn^{2+}, exchangeable Mn^{2+}, organically bound Mn and as various Mn minerals. The equilibrium among these forms determines Mn availability to plants. For satisfactory Mn nutrition of crops, solution and exchangeable Mn should be 2 to 3ppm and 0.2 to 5 ppm respectively. The continuous recycling of organic matter significantly contributes to soluble Mn. Mn soil fractions are Mn^{2+} and Manganese oxides in which Mn is present in trivalent or tetravalent form. Divalent Mn (Mn^{2+}) is adsorbed to clay minerals and organic matter and also the most important Mn form in soil solution. Mn^{2+} ion decreases 100 – fold for each unit increase in pH under high soil pH conditions. Manganese availability can thus be inadequate to meet plant demand. Liming decreases Mn availability. Plants take up Mn^{2+} which moves to the root surface by diffusion. Mn availability is higher in acid soils due to higher solubility of Mn compounds under low pH conditions. Mn^{2+} level of soil solution of acid and neutral soils is in the range of 10^{-6} to 10^{-4}M. In soil solution, Mn is present largely as organic complexes. Under dry conditions, Mn becomes less available. Mn^{2+} is fairly mobile in the soil and can be easily leached. Soil reduction increases Mn^{2+} and this content increases in saturated soils also. Acidity and lack of oxygen causes a, deluge of dissolved Mn^{2+}. Alkalinity and aeration favour conversion to less soluble Mn^{3+} and Mn^{4+} form. The DTPA extractant has been found successful for several crops on a variety of soils. The threshold value of Mn is 2.0 ppm. The critical levels are not absolute values and are likely to change depending upon the soil characteristics, the crop etc. However, the available content near approximation helps to identify the deficient soils.

4.2.10.2. Factors affecting Mn availability

1. pH regulates availability of Mn in soils. Increasing pH decreases the Mn availability. Mn availability is higher in acid soils. Use of acid fertilizers on calcareous soils produce beneficial effect on Mn availability.
2. Soil water logging reduces O_2 availability and lower the redox potential, which increases soluble Mn^{2+} in acid soils. Poor aeration and accumulation of CO_2 around root increases Mn availability.
3. Sandy soils in general, and acid sands in particular, are low in total Mn and will become Mn deficient if limed excessively.

4. Mn availability is low in soils with organic matter. Manganese forms unavailable compound with organic matter.
5. High Fe availability in soils may reduce Mn availability. For optimum plant growth, Fe: Mn ratios in the nutrient medium should be between 1.5 and 2.5. Ratios above 2.5 leads to Mn deficiency, while those below 1.5 cause toxicity. Similar to Fe, higher levels of Cu or Fe also reduces Mn availability in soils. Application of KCl to acid soils also increase the Mn availability and its concentration in plants. The effect of KCl on Mn uptake can be strong that it may produce even Mn toxicity in sensitive crops.
6. Wet weather favours the availability of more Mn^{2+}. Dry weather either induces or aggravates Mn deficiency in fruit trees. High light intensity as well as increasing soil temperature improves Mn uptake. Plants show a poor capacity to utilize soil Mn in cold weather with low light intensity.
7. Mn availability also decreases due to oxidization of Mn^{2+} to Mn^{4+} by certain soil bacteria and fungi.
8. Irrigation with water that increases the soil pH and deposits of bicarbonates and carbonates also aggravates Mn deficiency.
9. Several crop species exhibit differences in sensitivity to Mn deficiency. In general, fruit trees are usually considered to be the most affected by inadequate Mn nutritions.

4.2.10.3 Manganese in plants

Plants absorb manganese in the form of Mn^{2+} ion and the Mn concentration in plant tissues vary much more than any other micronutrient. Mn content in plants ranges from 20 to 500ppm. The manganese concentration in plants tend to decrease with age and to increase from lower to upper leaves. The critical concentration varies from crops to crops. It appears that Mn contents below 20 ppm are suggestive of its deficiency in several crop plants. Manganese toxicity is suspected if plants contain above 500 ppm, Iron content in plant also affects Mn centencatration in plant tissue. Very high or very low Fe content induces Mn deficiency or toxicity in plants. With respect to uptake of Mn, Mn uptake is generally lower than for other divalent cations. Manganese participates in cations competition. Magnesium in particular depresses Mn uptake. Manganese is relatively immobile in the plant. Manganese is mainly transported as Mn^{2+} and not as an organic complex. Manganese is preferentially translocated to meristematic tissues and hence young plant tissues are rich in Mn. Demand is highest at the time of maximal meristem development, and manganese levels are therefore high in organs that have a high level of metabolic activity. The manganese content of a plant generally increases as the plant develops. It has been reported that little manganese applied to leaves move to the other parts of the plant, and manganese deficiencies can therefore can be corrected by repeated foliar sprays. Manganese mobility is slightly higher in monocot crops than the dicot crops. Hence, the deficiency of Mn occurs in middle and some what older leaves in monocotyledons; whereas in dicotyledon crops, the deficiency occurs in younger leaves. The uptake of Mn^{2+} cations and their transport within the plant tissues are affected by high Ca content in plants. A Ca/ Mn ratio of about 350 is considered normal in plants.

4.2.10.4 Physiological role of Mn in plants

Manganese is important in plant metabolism because of its redox properties and thus involved in oxidation - reduction of carbohydrate and protein metabolism. It's high redox potential is regarded as the most important metabolic activator of many enzymes viz. pyruric acid dehydrogenase, á – Keto glutaric acid dehydrogenase, oxaloacetic acid dehydrogenase and oxalosuccinic acid dehydrogenase and so is particularly important for respiration in plants. Manganese deficiency therefore results in accumulation of citric acid. Some non-specific enzymes activated by Mn^{2+} include various phosphatases, phosphokinases, phosphate transferases and the enolases. Manganese activates phospholipase, the enzyme catalyzing the synthesis of phosphatidic acid which is the precursor of lipid viz., lecithin and cephalin. These two are indispensable for the synthesis of cell membrane including those of chloroplast and believed to contribute to the structural strength of chloroplast by its binding to their lamella. Thus, lack of Mn reduces the level of both green and yellow pigments.

Manganese is directly involved in the photosynthesis through its role in Hill reactions. Manganese increases the unsaturated fatty acids in plants. Manganese also plays an important role in protein metabolism. Plants lacking manganese accumulate nitrate and nitrite and disturb protein metabolism. Manganese is also involved in the hormonal regulation (IAA), Manganese detoxifies H_2O_2 in cell metabolism by activating peroxidase and decreasing cetalase activity. Plants receiving sufficient manganese are reported to need less water than those that are deficient. Therefore, Manganese is involved in the control of water status of the plant. Sufficient Mn content in plants induce early flowering in plants. Manganese also increases the cold tolerance in plants.

4.2.10.5 Diagnosis of Deficiency

Visual symptoms: The deficiency of manganese can be detected by appearance of visual symptoms mostly in leaves. This is a simple and inexpensive technique of diagnosing the deficiency. Field experience with high expertise can alone pinpoint the deficiency very easily. Manganese deficiency resembles Mg deficiency, as in both cases interveinal chlorosis occurs in the leaves. In contrast to Mg deficiency, however, Mn deficiency symptoms are first visible in the younger leaves; whereas in Mg deficiency, the older leaves are first affected. Manganese deficiency symptoms in the dicots are often characterized by small yellow spot on the leaves.

Leaf or plant manganese content serve as reflections on the available Mn status of soils, since it is an integrated effect of several interacting factors. The tissue Mn concentration also serve as reliable index in diagnosing Mn deficiency. Generally 20-30 leaves of youngest fully matured leaves just prior to flowering can be used as index tissue for analyzing nutrient content. Leaf or plant Mn content has to be interpreted based on the critical content of Mn which has been specified for each crops for deficiency and sufficiency.

Table 15: Critical content of Mn in fruit crops

Crop	Critical Limit (ppm)
Apple	15.0
Avocado	12.0
Citrus	18.0
Peach	10.0
Plum	15.0
Pear	25.0
Cherry	44.0
Mango	60.0
Banana	20.0
Grapes	35.0
Pomegranate	30.0
Walnut	20.0

Owing to the low mobility of Mn, it does not initially affect the whole leaf, but first appear in the mesophyll of the intercostals parts of the leaves at some distance from the veins. The chlorophyll is broken down in Mn deficient cells, loose their green colour and thus turn pale or yellowish green to form dot like chlorosis between the small veins. They appear as bright spots when the leaf is held against the light. Chlorosis and necrotic symptoms occur in fully developed some what older leaves. New leaves are often quite normal. In acute deficiency, even the youngest leaves become light to olive green.

Apple

The deficiency causes the leaves to turn yellowish-green sometimes-becoming pale yellow, while the main veins continue to remain green. These faded areas may be surrounded by deeper green areas. The leaves on the tips and the middle of the shoots are generally affected by the deficiency. Leaf size is not reduced but the crop yield can be reduced. The yellowing of the leaves is very similar to the iron deficiency symptoms. These can be distringuished in the early stages by the fact that the iron deficiency symptoms usually appear on the youngest leaves first; whereas Mn deficiency occurs in the tip and mid-shoot older leaves. Secondly , the normal green colour bands may be wider and more clearly defined from the chlorotic zones on manganese deficient leaves than on the iron deficient leaves. The Mn deficiency is more pronounced in dry summers and in dry areas.

Banana

The younger leaves show marginal interveinal chlorosis leading to coalescent necrotic spotting and finally a necrotic brown leaf margin. There is also some black-pimple spotting of the fruit.

Cherry

Leaves show interveinal chlorosis which may spread from margins over the whole leaf; leaves appear limp and chlorosis may be fairly general over whole tree.

Grapes

The deficiency is visible in the early part of the summer. The leaves at the base of the shoot start to pale, and shortly after wards small, polygonal yellow spots appear in the interveinal tissue. These are arranged in a mosaic pattern bordered by small green coloured veins. During drought, the margin of older leaves dry up while the younger leaves at the tip of shoots and on laterals remain green. The symptoms are more severe on sun exposed leaves than on shaded leaves. The growth of shoots, leaves and berries are affected by Mn deficiency and the maturation of the clusters is delayed.

Cashewnut

Interveinal chlorosis of younger leaves is the first visual symptoms of Mn deficiency. In some cases, brownish patches can also be observed near leaf margins. Leaf development will also be poor. In severely affected plants, chlorosis may cover the entire lamina except the major veins and laterals more conspicuous. Dwarfing and reduced leaf production are also due to deficiency of Mn.

Citrus

Manganese deficiency shows a light green mottle on a dark green background. Often the pattern appears like a series of inverted "horseshoes" around the periphery of the leaf. At times, it may be a complete marbling on light green on darkgreen background. Green-veins are affected due to either lacking or appear fuzzy in contrast to the sharp delineation of Fe deficiency. Leaf size is not reduced. If the stages of deficiency continues, the pattern persists upto maturity and old age of the leaf until every leaf on the tree is patterned.

Pineapple

A light yellowing of the plant is noted in the later stages of growth near harvest time but no chlorotic condition of the leaves develop.

Coconut

The palms are slightly discoloured but main symptom is narrow, longitudinal brown necrosis on leaflets parallel to the veins and their margins. By contraction, the necrosis induces curling of the leaflets. Cracks appear on the cortex of the primary roots.

Peach

Symptoms of manganese deficiency are chlorotic areas between the leaf veins, checkered yellow and green or sometimes even solid greyish yellow except for the green along the veins. In the latter stages reddish blotches or brown spots appear. There may also be drying and dying back of branches.

Avocado

Chlorotic areas develop between leaf veins, usually starting near midrib, developing up and down the leaf and gradually spreading out to the margins as the deficiency becomes more pronounced.

Pears

The deficiency causes chlorosis of the leaf tissue, beginning at the leaf margin and progressing inward. Tissues adjacent to leaf veins and margins turn green, producing a pattern similar to that caused by Fe deficiency.

Plum

In Leaves, interveinal chlorosis, begin at margins and spread over to the entire leaves on terminal shoots.

Strawberry

Leaves faint, interveinal chlorosis begin at the margins and later on spreads to the whole leaf.

Mango

Symptoms appear on one month old younger leaves. Leaves show yellowish green background with a fine network of greenveins, this pattern being marked on the upper surfaces and disappearing after a few weeks. Mature leaves are thicker and more blunted.

Raspberries

The leaves may show interveinal chlorosis over a considerable length of the cane. Diffused V shaped interveinal chlorosis also occur.

Walnut

The chlorosis develops between the main lateral veins and extends from the midrib to the margin. The deficiency begins to show in early to mid summer. The chlorotic pattern produces a "herring-bone" effect. The chlorosis gives an overall pale or slightly yellow appearance. Generally, leaf size is unaffected unless the Mn deficiency is very severe, in which case leaf size could be smaller and the yield is reduced.

4.2.10.6 Amelioration technique

Manganese deficiencies can be corrected by soil or foliar application of manganese carriers. Owing to the poor mobility of Mn^{2+} cations in the plant, two or three applications will be required, which will naturally cause additional cost for spraying unless the operation can be combined with crop protection chemicals or with the application of growth regulators. Light to moderate Mn deficiencies can be remedied by foliar applications of Mn containing pesticide such as Dithane (80 % Mancoceb).

Among the different sources, manganese sulphate ($Mn\ SO_4 . 3H_2O$) is the most widely used sources of Mn which contain 26-28 per cent and it is water soluble. Soil application @ 25kg/ha along with 12.5 t FYM through band application is recommended. In the case of foliar spray, spraying of 0.5-1.0 per cent manganese sulphate @ 400 lit/ spray fluid is found to be the best. Two to three sprays at 14 days interveinal are given depending upon the severity of the deficiency. A combination of soil application followed by foliar spray is recommended for severe cases on Mn deficiency of fruit trees.

4.2.11 Boron

Boron is considered as the one of the most important micronutrient next to zinc for growth and productivity of fruit crops. As early as in 1910 Agulhon observed that boron stimulated the growth of radish, oat and wheat and in 1923 warington proved the essentiality of boron for plant growth. Previously not much importance was given to boron, But now-a-days, its deficiency has been wide spread noticed in many parts of the country. Among the cultivated crops, the horticultural corps are susceptible to boron deficiency in general and tree fruit crops are in particular. It is a well-known fact that the dicotyledons crops require relatively higher boron than the monocotyledons crops. Boron is neither an enzyme constituent or catalyst for any enzyme. However, it is involved in vital physiological functions of plant through its binding with metabolites. Early diagnosis of the deficiency symptoms is of practical importance to prevent further losses by application of appropriate amount of the deficient nutrient.

4.2.11.1 Boron in soils

Boron is a member of metalloid group of elements like silicon and germanium. The chief sources of boron in soil is the mineral tourmaline. These elements are intermediate in properties between metals and nonmetals. Boron has three valancies. Boric acid is a very weak acid and in aqueous solution at pH <7, it occurs mainly as undissociated boric acid; at higher pH, boric acid accepts hydroxyl ions from water and forms a tetrahedral borate ions.

$$B(OH)_3 + H_2O \rightarrow B(OH)_4^- + H^+$$

Boron is absorbed as $B(OH)_3$ and exist in a soil solution as an undissociated molecules. The total boron content in soils ranges from 2 to 100ppm. The lowest boron content was found in soils derived from igneous rocks and fresh rudimentary deposits (coarse textured) and low in organic matter soils; whereas the soil derived from shale, loess and aluminum (fine textured)show higher boron. Boron is prone to leaching. Soils of arid and semi-arid regions contain higher boron than soils in humid zones. As sea water contains appreciable boron (4.7 ppm), soils under marine influence may also be higher in boron. Hot water soluble boron in soils is considered as available to plant. The critical limit for available boron in soil (Hot water) is 0.5 ppm.

4.2.11.2 Conditions favourable for boron deficiency

1. pH:Boron availability decreases as the pH increases particularly in calcareous soils as a result of the formation of $B(OH)_4^-$.
2. Higher clay content decreases the boron availability.
3. Calcareous soils may be deficient in boron because of their higher pH. Liming is also known to decrease boron availability.

4. Organic matter content increases the boron availability by
   i. Preventing its leaching loss
   ii. Brings about its accumulation in surface soils.
   iii. Moderating the ill effect of rising pH.
5. Texture: Coarse textured soils are inherently low in available boron. It's availability is low in sandy soils of high rainfall areas where boron is likely to leach down into the profile.
6. Soil moisture: Moisture conditions are regarded as conducive for boron availability. So, drought condition induces boron deficiency.
7. Light intensity: High light intensity induces boron deficiency.
8. Calcium content: Higher calcium content of soil promote boron deficiency in low boron soils. Irrigation on water with high calcium content also induce boron deficiency. Calcium antagonistic with boron.
9. Prolonged cropping without proper supplementary application of boron or inadequate supply of organic matter may also causes boron deficiency.
10. Crop sensitivity: Crops differ considerably in their boron requirement. Cereals require less boron and thus they are less susceptible to low boron in soils; whereas the leguminous and cruciferous crops have higher boron requirement and suffer more frequently from boron deficiency.

4.2.11.3 Boron in plants

Plant absorb boron mainly passively as boric acid $B(OH)_3$ but to some extent as borate ions $B(OH)^-_4$ and it is transported through xylem with the transpiration flow to the above ground parts like Ca^{2+} ions. The uptake of Boron and its transport in the plant is therefore closely linked with the rate of water uptake and transport through the xylem in to the different organs.

Boron concentrations vary widely among the plant species. Monocotyledons invariably contains less B than dicotyledons. Boron is relatively immobile in plants and the B content increases from lower to the upper plant parts. Hence, leaves contain more B than stems. Usually high B concentration are found in leaf margins. If plants contains less than 15 ppm B, they are suspected to suffer from B deficiency (the critical content may be lower than that for monocots). In cases where B content exceeds 200ppm, B toxicity is to be expected. The B content of the whole plant may decrease during the initial growth, then remain fairly constant during most of the vegetative phase. Higher concentrations of B occur plant organs such as anthers, stigma and ovary, where levels may be twice as high as in stems. In cases of deficiency, the apical part of plants have a lower boron content than the basal part; whereas in plants receiving adequate supples, the boron is uniformly distributed throughout the plant or is at the most slightly higher in the apical part. A good boron status also has a positive effect on phosphorus uptake. Increasing Ca content decreases the soluble B content or induces B deficiency. The demand for boron by the plants is highest when growth in the weight of leaves is highest at blossoming and when the fruits are set. Boron does not undergo any valency changes in plants.

4.2.11.4 Role of boron in metabolism

Boron is required for proper development and differentiation of tissues. It involved in many vital physiological activities of the plant. Boron forms complex with sugar for easy transportation through cellular membrane.

Boron increases the cell wall thickness by forming complexes with cell wall constituents and also by borate-ester cross-links.

Boron is involved in the cell elongation rather than cell division.

It is involved in RNA synthesis through N-base utilization.

It is also involved in the synthesis of protein.

It inhibits the IAA oxidation and there by decreases the IAA activation.

It has been proved that boron is essential for membrane formation and also for the membrane integrity.

Boron is involved in the germination of pollen as well as for the growth of the pollen tube. High boron levels in the stigma and styles are required for physiological inactivation of callous from the pollen tube walls by the formation of borate – callose complexes. This will facilitate easy pollen tube growth. Boron increases the pollen producing capacity of anthers and pollen grains viability.

In addition to this, boron also involved in lignin synthesis, starch formation, phenol metabolism, tissue differentiation and respiration.

4.2.11.5 Diagnostic Technique

Boron deficiency can be diagnosed by a) soil analysis, b) plant analysis and c) visual symptoms,.

a) Soil Analysis

Available boron in soil is extracted by hot water method. The available soil boron content less than 0.5ppm is considered as critical concentration and these soils are deficient in boron. Available boron content above 5.00ppm are commonly considered as high and one can expect toxicity in these soils.

b) Plant analysis

This can also be employed as a diagnostic tool to diagnose the boron deficiency. Different leaves or plant parts of the same crop differ considerably in their boron content. Therefore, the plant parts to be analysed for boron content should reflect the whole plant concentration. The middle leaves of new shoots (4 leaves per tree and 100 leaves per sample) have been standardized for analysis of boron content. The analysis will give clear picture about the concentration of boron. Based on the concentration, it can be easily judged for its deficiency or sufficiency. The critical concentrations for some of the fruit crops are given in table 16.

Table 16: Critical level of B in fruit crops

Crop	Boron content (ppm)
Peach	15.0
Plum	20.0
Apple	15.0
Pear	10.0
Citrus	15.0
Apricot	8.0
Grapes	10.0
Banana	10.0
Avocado	8.0
Guava	5.0
Coconut	14.0
Oil palm	8.0

c) Visual symptoms

The visual symptoms are the easiest method to diagnose the nutrient deficiency. Some times, the diagnosis based on the symptoms becomes uncertain, if the deficiencies of two or more nutrients occur simultaneously. It requires a special skill and constant touch with crops to diagnose deficiency symptoms of a particular nutrient. Visual symptom of nutrient deficiencies may be exhibited in all organs of the plant, viz. leaves, stems, flowers, fruits, seeds and roots. In the case of fruit trees, the symptoms on leaves and fruits are prominent.

Among the fruit trees, apple and pear are sensitive to boron deficiency; whereas fig, citrus and grapes are sensitive to boron toxicity. Boron toxicity rarely occurs in crops due to excessive application and also due to high boron content in irrigation water (more than 1ppm). Boron toxicity is expressed by yellowing of leaf tips. In extreme conditions, a progressive necrosis of leaf tissue extending between the lateral veins towards the midrib occurs. The tip and margins of leaves show a scorched appearance which later on covers the entire leaf before it sheds prematurely. Crops like, alfalfa, artichoke and corn are some of the indicator plants for boron toxicity. Liberal N applications have been useful in controlling the excess boron in many crops and thus alleviating the boron toxicity. Crops suffer from boron deficiency usually exhibit fairly characteristic symptoms. The first symptoms of boron deficiency develops when the level of boron in the soil or plant is just at or near the threshold of boron deficiency. Boron deficiency symptoms are often worse during or just after drought and it may be difficult to distinguish between the boron effect and the drought effect. The basic characteristic of boron deficiency which are found on many crop species are; the youngest leaves are first affected; they are misshapen, thick, brittle and small, but seldom exhibit any chlorosis. In fact often they are dark green. Stems are short; severely affected plants are liable to have a "shrunken" appearance. Growing points become moribund and die. Auxiliary meristem develop which will also show symptoms; plant becomes bush shaped. Necrotic and watery patches develop in storage tissue. Cracks and splits occur in petioles, stems and some times fruit. Fruit formation will be irregular as a result of incomplete fertilization. Fruits are likely to be misshapen. Leaves will tend to have a more simple shape. Root growth will be impaired. Whilst many of the symptoms of boron deficiency are very characteristic, conformation should be sought by xylem of plant or soil analysis. The visual symptoms of fruit crops are furnished below.

Almond

New shoots fail to develop from buds. Tip of branch dies in a few months, but lateral shoots farther down may grow, giving a brushy appearance. A large part of the tree may die. Brown gummy areas in nuts may extrude to surface; the nuts drop in May or June. Leaf tips from water sprouts scorch, curl and drop.

Fig

Terminal bud ceases to develop, then numerous lateral branches appear just back of tips. Leaves are chlorotic, necrotic at margins and between veins, and distorted. Internodes are short.

Blue berry

Leaves of terminal buds become bluish, young leaves are chlorotic in spots, later becoming blotched and distorted; older leaves are brownish.

Citrus

Wilting of trees is the first visible deficiency symptoms of boron. The wilting occurs in trees even under sufficient moisture status of the soil. Subsequent to this wilting, smaller water soaked spots in the leaves, becoming translucent appear as the leave mature. Sometimes, splitting and corking of the midrib and the veins of the leaf causing the leaf to curl downward. Such leaves are lusterless and are bronzed. Premature defoliation also occurs. In severe cases, multiple buds develop. The bark and shoot split in the internodal regions and gum exudates. The deficiency also causes break down of phloem tissue or internal girdling of young branches. Sometimes, malformation may also occurs. The fruit is mishappen with hard spots and gum deposits in the peel. The seeds are usually dark and under developed. The pulp and peel become hard and juice content is very low. The malformed fruits shedding are also very common.

Mango

Boron deficiency in mango casuses stunted growth and the internodes become short. The new terminal leaves are reduced in size, pale green, distorted and brittle. The older leaves are normal green but were smaller in size. Some leaves become curved on one side. The midrib is brown from the ventral side. Sometimes, the fruits are malformed and not smooth. Internally, corky areas may be seen and less in sweet. Generally boron deficiency is observed in sandy soils and also the orchards near to the brick factories.

Grapes

The first deficiency symptom of boron appears before bloom on the tendrills near the shoot tip and on stalks of inflorescence. Dark, knotty bulges form and become necrotic. The distal portions dry up and flower clusters die. During rapid shoot growth, young internodes swell slightly and darken at one or several places and the pith becomes necrotic. Older internodes develop swellings with deep folds and pits that often reach the pith. These swelling are the most characteristic of boron deficiency. Deformed internodes are thicker and shorter than normal ones. Boron deficiency in summer lead to apical necrosis of the shoot primordium in the dormant buds. In the following seasons, such damaged buds produce short, bushy, branched sterile shoots.

Boron deficiency on the leaves appears first as lighter areas in a mosaic pattern and the fine leaf veins in the interveinal areas turn brown. On the margin and between the veins, red brown spots appear and slowly darken. The main veins near them usually remain green. In deficiency, only a few seeded berries set and small, seedless berries develop called "hen and chicken" or the "pumpkin and pea" symptoms. In acute cases, the berries develop necrotic foci inside which account for their grey brown colour. Boron deficiency also affects the root growth. The root remain short, thickened, swell up in to knots and break open longitudinally.

Guava

The first visible symptom is the failure of elongation of the terminal bud. The younger leaves become pale green, loosing more chlorophyll at the base of the tip where the deficiency is acute. The leaves margin turn pink colour and it slowly spread over to the interior part of the leaves. Sometimes, the fruit may be cracking and roughening of the skin and pitting on the interior.

Papaya

Deficiency symptoms are visible first as light yellowing of the third, fourth and fifth leaves from the growing tip. The tip of the leaves curl downward with death of some tips and margins. The leaves are brittle and claw-like. In rapidly growing plants, there is latex exudation on the upper part of the tree trunk, leaf stalks and underside of the main leaf veins, which latter develop corky splits and cankers. After some period, the young leaves become markedly small with severe marginal distintion, brittle texture and necrotic margins. The deficient plants are stunted and may take some times to die. The root system is reduced in size with lampy roots and the terminal roots after dying. There is flower shedding in male trees. Fruit set will be poor in female trees. The fruit surfaces become rough and lampy in appearance. Most seeds in the deficient trees are abortive or poorly developed.

Banana

Boron deficiency results in the development of small chlorotic streak which are perpendicular to and crossing to the primary veins of the leaf blade. When the deficiency becomes severe, the chlorotic streaks become longer and more concentrated, eventually extending through the leaf. In certain cases, slight protrusions appear on the lower surface. As the deficiency becomes more acute, the interveinal chlorosis and leaf malformations occur. The leaves become narrow, rolled and incompletely developed. Sucker blast may also occur. In some cases, the blossom blast is also occur. Fruit set is also affected to a considerable extent.

Pineapple

Fruit are abnormal in size and shape. The fruit is also small with pronounced separation and cracking of fruit lets. The cracks between fruitlets are filled with gum. In sever cases, plants produce corcky cricket ball size fruits or no fruits.

Sapota

Boron deficiency in sapota is not observed in foliage. But its deficiency reduces the yield through decreased pollen tube growth, which results in low fertilization and fruit setting.

Pomegranate

Fruit cracking in pomegranate is due to the deficiency of boron.

Apple

Boron deficiency is recognized more easily on the fruit than on the foliage. The first symptom on the fruit is internal cork or brownish dead tissues inside the fruit which may appear any time shortly after bloom until harvest. The internal cork is round or irregular in shape with defenite margins. Immature fruits drop and may be enhanced in summer. External cork is usually developed before the fruit is half grown and develop as water soaked spots. These spots later turn light brown, wrinkle and exude amber coloured droplets, which soon hardened.

Deficiency symptom in leaves is manifested only under acute boron deficiency. The foliage turns yellow and red venation of the terminal shoot leaves, followed by marginal necrosis and tip necrosis. Death of small areas of bark near the shoot tip is followed by progressive death of most of the inner bark and cambium from the tip. The terminal bud fails to open and a portion of the terminal shoot die back. Below the point of death, lateral dormant buds may grow and die back again resulting in a dense bushy ' "Witches broom" growth, which appears to be susceptible to winter injury.

Peach

The boron deficient trees produce small, twisted, thick and leathery, malformed and yellowish leaves with corky main veins. Leaves may cup with some mottled yellowing. The dieback of twig, rosetting of new shoots, failure of leaf and bud development are also noticed in boron deficient trees. Necrosis of the leaf and stem, bark splitting and distorted bark. The symptoms are less similar to cold injury. Fruit shows gumminess, cork towards the center, blighted seeds and poor mealy teasteless in quality.

Apricot

The terminal twig die back leaves spatulate dwarfed curled at margins; necrotic at tips, narrow, brittle and chlorotic between veins. In the case of fruit, internal browing, formation of corky tissue, external cracking of fruit tissue, external browing are the important symptoms. The browing of tissue is more worse near the stone than the other sides. The affected trees produced the shriveled fruits.

Plum

The deficiency in fruit shows sunken areas in the flesh of fruits as a single small spot or may affect the whole fruit. The brown flesh beneath the sunken areas become firm in texture and in severe cases extended upto the pit. Fruits showing symptoms of boron deficiency usually develops colour earlier than normal fruits and drops. In some cases, gum pockets may be formed in the flesh of fruits. In the case of foliage, the dieback of twigs is the most common deficiency symptoms.

Pear

Reduced shoot growth with failure of basal leaves to develop. Affected twigs and severely affected trees may die during the winter season, Flower becomes brown in colour. Early leaf fall, Blossom of affected trees may fail to develop fruits causing a severe reduction in crops. In the case of fruits, shallow depressions on the fruit surface may develop which are most abundant at the calyx end of the fruit and may merge. The depression in the fruits are underlain in the flesh with corky tissues. In addition to this, cracking, splitting and malformation also occur.

Cherry

Boron deficiency in cherry tree leads to mild interveinal chlorosis of the young leaves associated with serration along the margins which are resinous and necrotic. Fruit are reduced in size with dark brown scars. Sometimes the splitting of fruits and sunken spots in the fruit flesh also occur.

Strawberry

The leaves were small, severely deformed and puckered with margins turning upwards. The tip of the leaves turned brown. The plant becomes less resistant to diseases. Reduced root growth also occurs. Sometimes, the death of terminal shoot point leads to the development of laterals, which also die back. The leaves become chlorotic and distorted while the petioles become brittle. The flowers burst and the fruits were malformed. Stolons were short with excessive branching and stubbiness in the roots.

Olive

The leaves are pale green and tip becomes yellowing. The terminal leaves become necrosis with a yellow-banded region between the leaf tip and the basal portion of the leaf. Leaf fall is very common. The boron deficient trees produce fruit, which are called 'monkey face' in olives. In these fruits, the endocarp growth remaining undisturbed, failure of mesocarp and exocarp in the terminal end of the fruit gives rise to a deformed fruit in which the endocarp protrudes through the apex of the fruit. Various patterns of flesh deformity are observed. Cracking and roughening of bark are also seen. The terminal buds often die without interruption of growth by shoots developing from lateral buds. This leads to 'witches broom' type of growth.

Raspberry

The boron deficiency in raspberry leads to die back which is characterized by delayed or complete failure of buds to develop on fruiting canes. A large number of buds die during the dormant season, while a few that sprouted produce distorted leaves with unusually long petioles. Later, the leaves become necrotic at the margins. In acute cases, the leaves remain small and had deeply indented leaflets that looked feathery.

Red currant

Boron deficiency in red currant leads to discolouration of foliage, shrivelled and blackened appearance of petiole of youngest leaf lets with neighboring leaves showing only a small green island in the centre of the leaf surrounded by wide dry, light brown band at the margin.

Walnut

Boron deficient trees are stunted with weak shoot growth characterized by short internodes, which gives the tree a bushy appearance. Sometimes, the deficient trees show many shoots, sharply curved at the tip, with rudimentary leaves or no leaves except at the tip. These affected shoots die in the next winter and form the so-called 'snake heads'. Leaves become curled and brittle with prominent veins. Leaves stalk showed splitting of cortical tissues. The twigs died back and the roots were dark brown, short and disintegrating. In extreme cases, there is serious dieback of shoots and branches.

Avocado

The boron deficient trees become short and the leaves turn yellowish-green in colour. The older leaves showed signs of burning at the tips and along the margins, which abscised prematurely. Abscission of leaves started from tips and continued towards the base of the shoot. The burning symptoms in the tip of the young leaves and splits or crack may develop on the midrib and large veins on the underside of the leaf. Acute deficiency leads to retardation of growth of the shoot and formation of axillary multiple buds which showed weak growth and finally died.

Cashewnut

Boron deficient trees fail to expand young terminal leaves which are distorted and remain rolled and the growing points die. The deficiency causes poor seed development. It's deficiency causes 'gumming' in nuts. The kernel quality is also reduced to considerable extent.

Coconut

Boron deficiency has been observed in both young and adult bearing coconut trees. In the case of young unbearing palms, palms had slightly withered and stunted apical leaf, crinkled leaves, hooked and fusion of leaflets; under acute deficiency, the palms had severely withered, stunted and dying apical leaf, crinkled, hooked and fusion leaflets and some fronds without leaflets.

Boron deficiency in adult palms causes malformation of leaves which are quite characteristic. The prominent symptom is the appearance of hook leaf, shortening of fronds and crowding of emerging leaves around the apex which gives a choked appearance. In some cases, the petiole of new leaf becomes very thick and forms a tubular structure enclosing the entire space at the apex. The boron deficient trees result in low productivity due to narrow and small nuts with lesser kernel content. Under acute case, barren nuts are also produced. Here the nuts will not contain any endocarp in kernel.

Oil palm

The emerging leaf shows lack of expansion. These leaves are usually wrinkled dark green and brittle. The development of pinnace is incomplete forming a tuft of deformed leaves. Boron deficiency caused abnormalities like 'Hooked leaf' with single or double hooks appearing on the pinnae, little leaf, fishbone leaf and blind leaf symptoms.

4.2.11.6 Remedial measures

Boron deficiency can be alleviated by judicious and timely application of boron containing fertilizers. Both soil and foliar applications are recommended depending upon the nature of the crop, soil and severity of the deficiency symptoms.

Generally, borax and sodium tetraborate are commonly used for soil application. Boric acid or solubor is suitable is suitable for foliar application. Colemanite is advantageous for sandy soils. Slow release boron frits are suited for sandy soils and also for high rainfall areas.

Boron may be applied to soil either broadcast or banded prior to sowing. The soil application may produce residual effect for the subsequent crops. The amount of boron applied for various crops ranges from 0.3 to 3.2 kg/ha depending upon the requirement and sensitivity of the crop to boron toxicity. Among the fruits, apple, pears, guava and grapes need regular attention of boron fertilization.

a). Soil Application

Boron may be applied to soil either broadcast or banded prior to sowing. The amount of boron applied for various crops ranges from 0.3 to 3.2 Kg/ha depending upon the requirement and sensitivity of the crop to boron toxicity. Among the fruit apple, pear, guava and grapes need regular application of boron fertilization.

b). Foliar Application

Foliar applications are quite successful for fruit crops. Three to four sprays may be given depending upon the severity of the symptoms. Boric acid or solubor at a concentration of 0.2 to 0.5 per cent may be suitable. Through wetting of 400 litres (per hectare) of spray solution mixed with 0.1% urea along with any wetting agent preferably in the new foliage facilitates the easy absorption. Boron can be combined with pesticides other than those formulated in oils and emulsions. Boron can also be combined with oils and emulsions. Boron can also be combined with other micronutrients other micronutrients.

4.2.12 Molybdenum

Molybdenum is required in smallest amounts among all the micronutrients. It's essentiality for higher plant was established as early as in 1939 by Arnon and Stout. Molybdenum is a transition element, which exhibits +4 and +6 oxidation states and forms of oxyanions like sulphur and phosphorus. Molybdenum deficiency is widespread in India particularly in Uttar Pradesh, Haryana, Madhya Pradesh and Gujarat.

4.2.12.1 Molybdenum in soils

The total Mo content in soils varies from 0.013 to 17.0 ppm with an average of 20.0 ppm. The available Mo content of Indian soils ranges between (Ammonium oxalate method) 0.01 and 1.65 ppm. Molybdenum exists as anion, MoO_4^{2-}) on the surface of soil particles and it is held less firmly and is considered easily available to the plant It is transported to plant roots by mass flow. Its deficiency causes considerable reduction not only in yield but also in quality of the produce.

4.2.12.2 Factors conductive for Mo deficiency

- Soils with the available Mo content less than 0.2 ppm (Ammonium oxalate method) is considered as deficient. Mo availability to the plant is influenced by many factors.
- The available Mo content in soils increases with clay content or fineness of soil texture. Sandy soils lose their Mo content more easily than heavy textured soils.
- Mo deficiency is found in soils with below pH 5.5.
- Soils with low organic matter are prone to Mo deficiency.
- Well-drained soils are found to be low in Mo availability.
- Excessive application of coppers aggrevates Mo deficiency.
- Soils with Manganese, Iron and Aluminum are found to induce Mo deficiency.
- In general, soils with low absolute Mo content are found in regions of high precipitation on slopes and also in soils with a coarse sandy texture where water percolates rapidly.

4.2.12.3 Molybdenum in plants

Molybdenum is absorbed actively by plant roots as ($Mo\ O_4^{2-}$). It is one of the moderately mobile elements in plant tissues. It is believed to have been transported through xylem vessels in the form of Mo amino acid complex, M-sugar and other ploy hydroxyl compounds. Its concentrations are highly varied among the crop plants and is dependent upon the species and variety. Mo requirement is higher in dicotyledons than the monocotyledons. Mo content in plant ranges between 0.1 ppm and 30 ppm. But healthy and adequately nourished plants contain 0.5 to 5.0 ppm of Mo. Roots contain more Mo than the shoots. It accumulates in the phloem and the vascular parenchyma. The midrib and margins of leaves are relatively lower in Mo content than the interveinal areas. Mo content in leaves and stems decreases at maturity with the accumulation in seeds. Among the crops, cauliflower, lettuce, spinach, tomatoes, beetroot and citrus species are sensitive to Mo deficiency; whereas, lucerne, cloves, rape, soyabeans, peas, carrot and celery are moderately sensitive to deficiency. Plants drawing their nitrogen from nitrate always need more molybdenum than those using ammonia and thus molybdenum requirement for plant growth is strongly dependent on the sources of nitrogen supplied to the plant.

4.2.12.4 Molybdenum in Crop physiology

Molybdenum is an essential component of various enzymes viz., aldehyde oxidase, hydrogenase, sulphite oxidase, nitrogenase and nitrate reductase. The aldehyde oxidase is found to reduce molecular nitrogen in bacteria. Nitrate reductase is an important enzyme in nitrogen metabolism. It catalyses the reduction of nitrate to nitrite. It is present as molybdeflavo protein in nitrate reductase. Nitrogenase is important for nitrogen fixing organism. It is present as Fe-Mo-S- complex in the ratio of 20:20:20: in one of the two-enzyme protein. It offers resistance to low temperature and water logging. Mo is important for germination of pollen grains.

It is essential for the synthesis of ascorbic acid. It is believed to enhance both the physiological potency and uptake of iron. It is considered an antidote of excessive Cu, B, Ni, Ca, Mn and Zn in plants. It is also thought to be indispensable for synthesis of nicotinic acid, which is an important co enzyme of dehydrogenase in energy metabolism. It is important for protein synthesis. It also acts as an activator of Xanthine oxidase and phosphatases.

4.2.12.5 Visual Symptoms

The crops exhibit the deficiency of Mo when the level in plant tissue is less than the critical concentration and an average of less than 0.1ppm is considered as critical. Due to its moderate mobility in plant tissue, the visible symptoms appear mostly on whole plant. In view of its metabolic functions in plants, it might be expected that Mo deficiency symptoms will resemble more or less that of nitrogen deficiency owing to the disturbance of nitrogen metabolism and the deficiency mainly affect the young leaves.

The symptoms usually appear first on the older leaves and then spread to those that are younger. The leaves wilt starting at the margins and accumulation of nitrates leads to leaf scorch which advances inwards from the margins. Chlorosis can also spread from margins to intercostal areas of young and old leaves. Usually irregularly oval shaped glassy chlorotic blotches appear in the intercostal leaf tissue. Damage to the meristematic tissue in leaves that are still green can lead to a slight thickening of the leaves and also to initially glassy, but later papery, thin, whitish brown necrotic blotches.

It restricted the vegetative growth and their leaves become pale and eventually wither. Flower formation may be restricted. Its deficiency appear first in the middle and older leaves. The yellowish green and the leaf margin roll in on themselves. Leaves are often small and covered with necrotic spots.

Apple

In younger leaves, there is a light greenish interveinal yellowing of the main veins while small veins remain green. The older leaves show a marginal burning due to the accumulation of nitrates under low Mo supply.

Citrus

The deficiency symptoms appear first as water-soaked areas on the leaves during the early spring, which coincides with new growth. Subsequently, these spots develop into larger interveinal yellow spot with gum on the lower side. In severe cases, leaves drop and in extreme cases complete defoliation may occur. Finally, it leads to die back of plants. Fruits show marked break down of the rind in several cases.

Plum

Deficiency of Mo causes a reduction in leaf size. Some leaves show a diffuse mottling and develop light brown areas of dead tissue at tips and margins.

Grapes

Yellowing and necrosis of leaf occurs in the absence of Mo.

Papaya

Mo deficiency cause significant reduction in growth in papaya. The old leaves develop mild interveinal chlorosis from the tip of the lobes. Later, the affected area becomes bleached, turned wavy and puckered.

4.2.12.6 Correction of deficiencies

Mo deficiency can be corrected by applying Mo fertilizers.Mo deficiency induced by strong acid condition can be corrected by liming the soil to pH above 5.5. It's deficiency can also be corrected by soil, foliar and seed treatment with Mo. Sodium molybdate and Ammonium molybdate are commonly used.

Seed pelleting with Mo at the rate of 50 to 100 g/ha in a liquid or slurry is recommended. Seed treatment avoids wastage of Mo and preventing deficiencies during early growth and establishment. As molybdenum is highly pholem-mobile, foliar spraying is appropriate and the deficiency can be corrected by foliar sprays of 0.1- 0.3 per cent of Mo salts.

Soil application of Mo varies for different crops and soils. Horticultural crops require a relatively higher Mo than the field crops. In general, a quantity of 200-400 g/ha of Mo is recommended. Single soil application may produce residual effect for 2-3 years. However, the crops which are sensitive to Mo deficiency need regular applications. Mo deficiency can be rectified by drenching the seed bed with solutions of Mo carriers. Two to three g of sodium molybdate in about 5-6 litres of water is enough to treat one square meter of seed bed.

Mo deficiency can also be avoided by sowing seeds with enriched Mo content. This is possible by foliar sprays of Mo during the seed development phase of mother crops. Application of Mo not only rectifies Mo deficiency but also helps in nitrogen economy.

CHAPTER 5

Management of Nutritional Disorders

5.1 Foliar nutrition of fruit crops

Foliar application of nutrients is gaining importance in horticultural crops in general and fruit crops in particular. The important criterion of the effectiveness of nutrient spray is the rate at which the foliar applied nutrients are absorbed by the leaves and translocated within the plant. For getting optimum yield, foliar fertilization should be carried out when the plant's requirements are particularly high and the nutrient uptake through the plant roots is restricted. The restriction may be due to either plant or soil limitations.

5.1.1 Situations needed for foliar fertilization

1. Foliar fertilization normally does not adapt itself to the application of large amounts of nutrients and should be considered as supplementary in a sound soil fertility programme.
2. Plant nutrients applied as foliar sprays are taken up through the leaves. Sometimes, soil conditions prevent crops taking micronutrients through the roots, during which foliar sprays are more effective.
3. The most common use of foliar fertilization has been to supply micronutrients to crops, particularly horticultural crops more specifically perennials.

5.1.2 Conditions determining the feasibility of foliar fertilization

The usefulness of foliar fertilization depends on the following conditions:

a) The existence of special problems that may not be coped with as well as by applications of the foliar to the soil or by soil management. It includes rapid fixation of the soil applied nutrients.

b) Response of plants to the nutrient spray.

c) Economic aspects and methods of application.

5.1.3 Factors affecting foliar absorption

i. Contact angle and surface absorption: The ultimate aim of the aqueous solutions of nutrients sprayed on leaves must be to penetrate living cells in order to be metabolized by them or translocated within the plant. The initial entry may be via stomatal apertures to the cells of the mesophyll as well as directly through the cuticle and into the cells beneath it. The ability of a liquid to wet solid surface is a function of its contact angle on the surface, this depends on the surface tension of the liquid and the nature of the solid surface. Addition of wetting agents resulted in reduction in surface tension, contact angle and increases the efficiency of absorption for the leaves.

ii. Path of entry: In the leaves, the pectinaceous materials, which should have greater water absorbing power, formed a continuous path from the layers in the cuticle, through the anticlinal walls of the epidermal cells to the cell wall and bundle sheaths surrounding the larger veins of the leaves.

iii. Age and nitrogen status of absorbing leaves: The leaves which are grown under high nitrogen conditions are more efficient in absorption of urea than low nitrogen in the leaves.

iv. Temperature and Humidity: Variations in temperature and vapour pressure deficit affect the rate of drying and the opportunities for establishment of a liquid film at the leaf surface.

v. Absorption of nutrient sprays takes place over considerable period of time and occurs when the leaf surface to be dry.

vi. Losses of nutrient sprays to the atmosphere and soil: Nutrient losses are taking place when nutrients are applied under following field conditions.

  a. Losses due to failure of the spray to reach of the leaf surface.

  b. Loss due to drop from the leaves.

  c. Loss due to volatilization

These losses vary greatly in relation to following circumstances

i. The degree to which the spray equipment used limits its application to the surfaces.

ii. The period of time that lapses between spraying and rains or dews which wash the leaves.

iii. The degree to which rainfall washes the leaves when it occurs.

iv. The adherence and solubility of the nutrient compound on the outer surface of the leaf.

5.1.4 Solution preparation

For spray solutions, fertilizers needed to obtain the desired concentration is given in Table , assuming that 400 liters of solution is required per hectare. This shows that at 0.5% concentration, one needs 2 kg fertilizer per hectare per spray.

Table 17: Fertilizer requirement for spray

Concentration of fertilizer in solution (%)	Grams of fertilizer in one litre	Kg fertilizer needed for a spray of 400 liters
0.5	5	2
1.0	10	4
1.5	15	6
2.0	20	8
2.5	25	10

5.1.5 Surfactants

Surfactants improve wetting of the leaf by lowering the surface tension and consequently reducing the contact angle between the spray fluid and leaf surface. Complete wetting is achieved when the contact angle is zero. The liquid surface tension at which the contact angle is zero is termed critical surface tension. Low surface tension will facilitate stomatal infiltration. Tween 20, sandowitch, teepol, ordinary soap solution can be used as surfactants. Generally, it is used at 0.01-0.3% (V/V). It has been observed that tween 20 (0.01%) doubled the absorption rate of urea in crops within 2-5 hours.

5.1.6 Time of spray

Crops are characterized by growth stages needing a particular supply of all or of certain nutrients to reach optimum yields. Crop demand is generally at its maximum during the period of exponential growth. During such critical stages, leaves show particularly high efficacy for absorbing nutrients. Further, in any vegetative period there are times when the nutrient uptake from the soil is impeded to such an extent that there is no of guarantee that all the parts of the plant will receive an adequate supply of nutrients. When there is a prolonged period of dry weather, even foliar fertilization is no longer able to exercise a favourable influence on the yield. The recommended timing of foliar sprays generally reflect growth stages of the single crops where the demand for specific nutrients is particularly critical (Table).

Table 18: Time of spray and concentration of foliar nutrition in fruit crops.

Critical nutrient	Recommended Growth stages	Concentration of nutrients for foliar spray
N	Before or after bloom, post harvest spray (leaves still in good condition)	1.0 % Urea
K	2,4 and 6 weeks aften bloom	1.0% Potassium sulphate
P	At flower initiation	0.6 % Potassium di hydrogen phosphate
Ca	3 weeks before harvest (Post harvest sprays)	1.0% Calcium ammonium nitrate
Mg	At petal fall, next sprays at 2 weeks intervals	1.0% Magnesium sulphate
Fe	4 weeks after bloom and 3 weeks later	1.0% Ferrous sulphate
Mn	At petal fall and 4 weeks later	0.5% Manganese sulphate
Zn	3-4 weeks after petal fall, post harvest sprays	0.5% Zinc sulphate + 0.25% lime
Cu	Immediately aften bud breaks	0.025% Copper sulphate
B	During bloom	0.25% Borax

5.1.7 pH of the nutrient solution

Phosphoric acid is absorbed more readily than the other P compounds. To some extent, K_2HPO_4 can also be used for foliar nutrition. However, H_3PO_4 and K_2HPO_4 should be used in low concentrations. $MgCl_2$ is more effective than $MgSO_4$ for foliar spray. However, $MgCl_2$ causes injury. The chelated – Fe are more effective than the inorganic salts due to the increased mobility of chelated Fe in plant tissue. Absorption of certain mineral nutrients by leaves and fruits have been found to be pH dependent. Absorption of urea by leaves is highest between pH 5.4 and 6.6, intermediate at 8.3. The optimum pH for Ca^{2+} absorption has been found to be 7. In the range of pH between 3 and 10, maximum P absorption occurred with sodium phosphate at pH 3 - 6 and potassium phosphate at pH 7 - 10. The pH had no effect on P absorption from ammonium phosphate or from calcium phosphate.

The addition of urea to the spray solution was reported to increase the effectiveness of foliar spray containing P, Mn, S, Mg and Fe in different species. However, addition of urea to the spray solution failed to increase uptake of Mg or P. Conversely, presence of Mg reduced the absorption of urea in leaves.

Generally, urea is readily absorbed by leaves and broken down by urease. The N is then incorporated into aminoacids and then to proteins. The proteins in turn are converted to amino acids, transported to the storage tissues, and reassembled into proteins. N absorbed by leaves during senescence is highly mobile in fruit trees.

Table 19: Absorption rates of foliar nutrition

S.No.	Nutrient	Time for 50% absorption
1.	Nitrogen	0.5 – 2 hours
2.	Phosphorus	5 - 10 hours
3.	Potassium	10 - 24 hours
4.	Calcium	10 - 94 hours
5.	Magnesium	10 - 24 hours
6.	Sulphate	5 - 10 days
7.	Chlorine	1 - 4 days
8.	Iron	10 – 20 days
9.	Manganese	1- 2 days
10.	Zinc	1 - 2 days
11.	Molybdenum	10 - 20 days

5.1.8 Suggestions for effective spray application

1. The deficiency symptoms diagnosed should be checked before spraying.
2. Do not exceed the recommended solution concentrations.
3. Adjust concentrations according to type of spray used (Particle size of the spray droplets)
4. Do not spray during hot, windy periods and against the breeze.
5. Wherever necessary, neutralize spray solution with lime.
6. Use only clear filtered solution
7. Repeat spray as and when recommended.
8. Avoid spray during mist to prevent washing away of the spray.
9. Double check the calculations used for preparing the final solution.
10. Do not expect a foliar spray to leave residual effect for next crop.
11. Do not store spray solution, particularly of iron. Use freshly prepared solutions.

12. Use wetting agents to improve effectiveness of spray.
13. Before buying a product, examine clearly its nutrient make up and composition and whether it meets the growing demand.

5.1.9 Merits of foliar application

1. Application rates are much lower than the soil application.
2. Uniform distribution is easily obtained.
3. Response to the applied nutrient is almost immediate. Therefore, deficiencies can be corrected during the growing season.
4. Suspected deficiencies can be more easily diagnosed with spray trials.

5. 1.10 Precautions during foliar fertilization

The hardness or amount of salts contained in the water, used to make the solutions affect the degree of concentrations of which the fertilizer chemicals may be applied with safety. The 'harder' the water, the greater the concentration of the fertilizer chemicals can be used. Hydrated lime can be used in soft water to enable the absorption of more concentrated solutions.

One must know the causes of deficiency and the proper chemical to use. Otherwise, additional injury may be caused by spray application. The following points are to be taken care of while planning for foliar spray.

1. Nutrient demand often is high, when plants are small and leaf surface is in sufficient to foliar application.
2. Leaf burn may result if salt concentration is in excessive.
3. It may be too late to correct the deficiency and still obtain maximum yield.
4. There is little residual effect from foliar sprays.
5. Extra costs may be required when more than one spray may be needed.
6. Boron cannot be applied with $CaCl_{2.}$
7. When combined with pesticide, the solution be acidified to lower the pH to about 6. High pH may be one of the reasons for occasional leaf injury caused by concentrated sprays of $CaCl_2$.
8. Copper deficiency can also be readily corrected with Bordeaux mixture.
9. Effectiveness of foliar sprayed Iron can be increased by addition of 0.1% citric acid (1% $FeSo_4$ + 0.1% citric acid).

10. Epsom salt sprays ($MgSO_4$) is compatible with most pesticides.

11. Spraying $MnSO_{4-}$ with fungicides Dithane M-45 is found very effective increasing leaf Mn levels.

12. Application of manganese sulphate to the soil to correct Mn deficiency is not effective on alkaline or heavy limed soils.

13. In the case of potassium, in general for fruit crops, the recommended doze is 0.84 Kg KNO_3 / 100 liters of water at 2.4 and 6 days after bloom increase fruits yield.

14. Urea should not be mixed with cyprex, Dodine, Karathane, Dikar etc.

15. Higher rates of Zn may cause severe injury to shoots, buds, leaves or fruits.

16. Addition of Mn to zinc foliar spray provides better response.

17. Addition of urea improves Zn absorption.

18. Foliar application of B, Cu, Mg, Mn, and Zn for controlling deficiencies of these elements in fruit trees have advantages over soil application. Those advantages are high effectiveness, rapid plant responses, convenience, elimination and reduction of toxic symptoms. The disadvantages of foliar nutrition are that the effects or sprays are temporary and are not transmitted in to the next year. Thus, the annual sprays are necessary.

19. Storage of $FeSO_4$ solution should be avoided otherwise ferrous Fe will be oxidised to ferric and precipitated.

20. Zinc sulphate can be mixed with phasolone or monocrotophos and $ZnSO_{4-}$ should not be mixed with phosphatic fertilizer.

21. Generally 400 litre /ha at spray fluid is recommended for hand operated sprays.

22. Ammonium sulphate is unsuitable for folior application. Because the solution is acidic and if the solution is neutralized by lime water, ammonia volatilises causing N loss.

23. If the concentration of fertilizer solution is high, it may cause salt burn. The concentration of N fertilizer, such as urea, usually does not exceed 2%. The concentration of H_2PO_4 should be very low i.e. 0.4% or less. Otherwise, it will injure the plants.

24. Sulphate salts are recommended for foliar spray at 0.5-1.0% and slightly vary depending upon the crop and local condition. Foliar spray of single

nutrients can be done depending upon the deficiency. Sometimes, same plant may exhibit deficiency of more than one element. In such cases, it does not mean that each compound can be added at 0.5% - 1.0% concentration. In such situation, total concentration should be within this level. Obviously, in such cases, the individual nutrient is applied at one third of the total concentration.

5.2 Chelates

The use of chelates in agriculture began in the early 1950's, with the work of Stewart and Leonard who demonstrated that soil application of Fe-EDTA to citrus growing on the acid infertile deep sandy soils of central Florida was more effective than $FeSO_4$ in combating Cu induced Fe chlorosis. The term 'chelate' is derived from chela, a greek word for a crab's claw. The chelates are also called siderophores. The chelates may be formed either naturally between organic matter and metal ion (natural chelates) or artificially between different chelating agents. The natural chelates are formed by reaction between cation and the organic substances come from exudates, humus decomposition, microbial cell exudates, animal manures and even polyphenols from leaf surfaces dissolved in rain water. When the organic substance bonds to the metal by two or more 'contacts', the organic substance is called a ligand. The ligand combines with metal forms the chelate. Metals in chelates are not removed from the organic compound. The ligand (organic radicle) which is negatively charged, forms covalent bonds with positively charged metal cation. The nutrients which can be chelated are Ca^{2+}, Mg^{2+}, Cu^{2+}, Fe^{2+}, Mn^{2+} and Zn^{2+}. The other micronutrients viz., B, Mo and Cl do not form chelates.

Chelates formed between synthetic chelating agents and cation are easily soluble in water. The chelating agents are known as sequestering agents. Some of the synthetic chelating agents are given in the table.

Table 20: Major chelating agents

Name	Formula	Abbreviation
Ethylene diamine tetra acetic acid	$C_{10}H_{16}O_8N_2$	ETDA
Diethylene triamine penta acetic acid	$C_{14}H_{23}O_{10}N_3$	DTPA
Cyclohexane diamine tetra acetic acid	$C_{14}H_{22}O_8N_2$	CDTA
Ethylene diaminedi (O-hydroxy phenylacetic acid)	$C_{18}H_{20}O_6N_2$	EDDHA
Hydroxy ethylethylene diaminetriaacetic acid	$C_{10}H_{18}O_7N_2$	HEDTA
Nitrilotriacetic acid	$C_6H_9O_6N$	NTA
Ethylene glycol-bis(2-aminoethyl ether) tetra acetic acid	$C_{14}H_{24}O_{10}N_2$	EGTA
Citric acid	$C_6H_8O_7$	CIT
Oxalic acid	$C_2H_2O_4$	OX
Pyrophosphoric acid	$H_4P_2O_7$	P_2O_7
Triphosphoric acid	$H_5P_3O_{10}$	P_3O_{10}

Among the several chelating agents, commercially used are EDTA, DTPA, CDTA and EDDHA. The chelating agents in general vary in their stability and suitability as sources of micronutrients.

5.2.1 Why chelates are effective?

If we apply micronutrients to soils, it suffers when it interacts with soils, for clays have negative surface charges that strongly attract and immobilize the positively charged metal ions. Thus clay effectively lock up the micronutrient. In a chelated form, however, the negative charge of the chelated metal micronutrient allows movement of the micronutrient in the soil solution for rapid, effective and efficient uptake by root system. The chelate is unaffected by the charges on the clay surface. When chelated nutrients are applied to the soil, it diffuse to the root cell where metal ions are absorbed leaving the chelating agent to rediffuse and mobilize further metal ions from the solid phases of soil.

5.2.2 Stability of chelates

The chelates are very stable. The stability of a chelate is affected by number of rings formed, the metal ion and the solution pH. Chelate stability also varies depending on the specific chelating agent used. For four micronutrients, the following order of stability of some of the major chelates has been calculated by Norvell (1972).

Fe: EDDHA > DTPA, CDTA > EDTA > EGTA = HEDTA > NTA

Cu: DTPA > HEDTA = CDTA > EDTA = EDDHA > EGTA = NTA

Zn: DPTA > CDTA = HEDTA = EDTA > NTA > EGTA

Mn: DPTA = CDTA > EDTA = EGTA = HEDTA > NTA

Generally, bonds are stronger with metals in approximately in this order:

$Fe^{+3} > Cu^{2+} > Zn^{2+} > Fe^{2+} > Mn^{2+} > Ca^{2+} = Mg^{2+}$

Except in soils of very high pH, iron chelates tend to be more stable than those of copper and zinc, which are inturn, more stable than those of manganese. Hence, iron is more strongly affected by the chelating agents than the other micronutrients.

5.2.3 Methods of application

Chelates can be used in several ways to correct nutrient deficiencies. They can be added as a spray on the foliage or they can be added to the soil. Different kinds of chelating agents are used. Some are effective on acid soils, some on calcareous soils, and some are effective on both acid and basic soils.

5.2.4 Characteristics of a chelate applied as foliar spray

1. Easily absorbed by plants.
2. Translocated readily within the plant.
3. Easily decomposed so that the metal becomes available.
4. Non damaging to plants at concentrations necessary to control deficiencies

5.2.5 Characteristics of a chelate applied to soil

1. Should not be easily replaced by other polyvalent cations in the soil
2. Must be stable against hydolysis
3. Has to be resistant to microbiological decomposition
4. Should be soluble in water
5. Not easily precipitated by ions or colloids in soils
6. Must be available to plants either at root surface or within the plant.
7. Must be non damaging to plants at concentrations required to prevent deficiencies.

Iron chelates*:* These are the most widely used. Ferric iron forms the most stable chelate. Therefore, it will remain in the soil in an effective form. EDDHA is the effective soil treatment for iron chlorosis.

Copper chelates*:* Next to ferric iron, copper chelates are the most stable and should be an effective source of copper. They have not been used commercially because copper can be utilized by plants from copper salts, even when the pH is high, because of the strong chelating action of soil organic matter for the copper ion supplied as a salt.

Zinc chelates*:* Soil application of zinc chelates have been more effective than the stability constant would indicate. The success is partly due to the fact that zinc chelates do not become fixed on the soil clay as iron chelates do and partly due to the low concentration of ferrous (0.000027 ppm) and ferric (0.00000006 ppm) ions at pH 6.0. At higher values, concentration of ferrous and ferric ions decrease rapidly. It is at soil pH values of 6.0 and above that zinc deficiency usually occurs. The use of zinc chelates will probably be limited because less expensive zinc sprays on foliage provide satisfactory control of zinc deficiency.

Manganese chelates*:* Because of the low stability constant, soil application of manganese chelates are not effective. Calcium exists in soil at high concentrations and by mass action will replace some of the manganese from the chelate even though calcium is lower than manganese on the scale of stability and replacing power.

5.2.6 Practical considerations of chelates application

- Chelated micronutrients can be mixed with other inorganic NPK and other fertilizers.
- Albio is a aminoacid chelate. Since aminoacids are the basic building block of protein formed in all living things, the chelation of minerals with aminoacid provides a tremendous advantage in the efficiency of absorption and translocation of minerals within plants. These chelated minerals are marketed under the trade name metalosated.
- Chelated micronutrients are costlier than the ordinary inorganic micronutrient salts.
- Chelated nutrients are most effective, smaller amounts are required to correct the deficiency than the inorganic micronutrient salts.
- Chelated nutrients should not be mixed with liquid fertilizers that contain ammonia. Due to high pH of ammonia, the metal forms precipitates because the (OH) groups compete with the donor groups of the chelating agent for metal ions.

5.3 Fertigation

Fertigation is feeding of crop by injecting soluble fertilizers into water in the irrigation system. It is also defined as precise application of water soluble fertilizers in the desired concentrations or fertilizer application through drip irrigation.

5.3.1 Types of fertilizers for fertigation

Fertilizers required for fertigation are of two types, namely (a) Water soluble fertilizers (b) Liquid fertilizers

a). Water soluble fertilizer (WSF)

The water soluble fertilizer are in solid form but completely soluble in water, carrying two or more major as well as micronutrients.

b). Liquid fertilizer (LF)

Liquid fertilizers are solutions which contain one or more nutrients in liquid formulation or in suspension form. Liquid fertilizers are pure, precipitate free and normally acidic (pH 5.5 – 6.5) which help in correction of soil pH.

When using fertilizers, which make a clear solution, a coloring dye can be used to show that nutrients are being added. Technical grade dyes are used such as EOCIN or FLUORECIN SODIU M (URANIN) @ 1g/ lit.

5.3.2 Nutrient sources

A variety of fertilizers can be injected into drip irrigation systems.

a. N sources : Urea, ammonium nitrate, calcium nitrate and potassium nitrate.

b. K sources : Potassium chloride, Potassium sulphate, potassium thiosulphate and potassium nitrate.

c. P sources : The choice of phosphorous products is limited. Phosphoric acid or ammonium phosphate is used most commonly. Monoammonium or Monopotassium phosphates are available, but not used frequently

5.3.3 Application methods

Fertigation can be applied efficiently utilizing any one of the following micro irrigation systems.

a) Drip irrigation : Water and nutrients are supplied simultaneously.

b) Sprinkler irrigation : Injection of fertilizers into sprinkler systems using aluminum plastic pipes is extremely effective.

c) Jet irrigation : Can be used for fertigation in orchard plantations of mango, pomegranate, guava, banana, etc.

Irrigation systems followed in India are mainly flooding method which not only results in huge loss of water but also costly. The use efficiency of both water and fertilizers could be increased substantially through drip irrigation and fertigation. Fertigation is the process wherein fertilizer is applied through an efficient irrigation systems like drip. Fertigation enjoys various advantages like higher use efficiency of both water and fertilizer, minimum losses of nutrients due to prevention of leaching , optimization of the nutrient balance by supplying nutrient directly to root zone in available forms, control of nutrient concentrations in soil solution to proper supply, saving in application cost, improvement of soil physical and biological conditions due to proper maintenance of soil moisture levels and so on. In fertigation, nitrogen use efficiency could be as high as 90% compared to 40 – 60% in conventional methods.

The advantages of fertigation indicate that nitrogenous fertilizer used in fertilization is used very efficiently by drip irrigation. Small amount of fertilizers are given at regular intervals rather than giving one big dose at planting. This means the amount of fertilizer, which is lost through leaching, can be as low as 10 percent. In the traditional systems, 50 percent loss is quit inevitable. The efficiency will be much more when phosphate and potassic fertilizers are combined in fertigation.

The other advantage of fertigation is that as the crop growth advances, we can change the formula of the fertilizers. During vegetative stage, the plant might need more of potassium. During the early growth stage, plant needs more of phosphorus. So according to the requirement, the ratio of fertilizer ca be easily be adjusted in drip system. It is not possible to do with the standard fertilizers applied in one or two applications during the cropping season.

Fertigation through drip is ideally suited for application of soluble fertilizers. Combined application of water and fertilizer is ideal for proper crop growth with irrigation water acting as a carrier for the nutrients required by crops. In this way the fertilizer is conveyed directly to the rhizsosphere through frequent application of soluble fertilizers in small quantities through the irrigation systems. Under proper management, this method opens up new avenues for growing crops under conditions similar to those of nutrient solution. This approach reduces nutrient losses, as a result fertilizer use efficiency is higher. Fertigation facilitates application of soluble fertilizers in many splits.

5.3.4 Suitability of nutrients

With regard to the suitability of individual nutrients for fertigation, nitrogen is the most common nutrient element applied through drip fertigation. Generally, all the nitrogen fertilizers are suitable for drip fertigation since they cause little clogging and precipitation problems except ammonium sulphate which may cause precipitation of $CaSO_4$ in hard calcium rich water. Urea is well suited for drip fertigation.

Phosphorus is not usually recommended for application through drip irrigation systems because of its precipitation as phosphate salts particularly when the quality of irrigation water is poor. No traditional phosphate fertilizer is suitable for fertigation. Phosphoric acid is the best source for phosphorus fertigation.

Application of the potassium fertilizers does not cause any precipitation of salts except in the case of K_2SO_4 with irrigation water containing high amount of calcium. Common potassium sources are readily soluble in water.

Micro nutrients such as iron, manganese, zinc and copper can be applied through irrigation water as chelated form (eg. FeEDTA), without causing any precipitation problem.

A correct rate and concentration of fertilizer is desired in fertigation system to avoid over fertilization and achieve the best results. It is to be specifically worked out for different cropping situations.

5.3.5 Fertigation equipment

A fertilizer tank is provided at the head of the drip irrigation systems for applying fertilizers in solution directly to the field along with the irrigation water. Soluble chemical fertilizers can be injected into the drip system and applied to the crop. The equipment for fertilizer application is relatively cheap and simple and can be fabricated locally. The fertilizer application consists of a sealed tank with necessary tubing's and connections. A venturi fixed in the main line creates the differential pressure and allows the fertilizer solution to flow in the main water line. Fertilizer is also introduced into the system from the suction side of the pump through a pipe and regulated by a valve. Another pipe is connected from the discharge side of the pump to the fertilizer container from the required water supply in the tank. This system is relatively simpler, but impellers are likely to be corroded due to the fertilizer solution, unless the impellers are made of corrosion resistant materials.

5.3.6 Preparation of solutions

Fertilizer application requires the maintenance of a constant concentration level which may be expressed in percentage or in ppm.

1% solution = 10,000 ppm

(10g/1 lit) = 1% = 10,000 ppm

5.3.7 Prerequisites for fertigation

i. **Soil water status:** The nutrient status should be assessed before finalising fertilizer dosage and frequency of injection.

ii. **Crop nutrient requirement:** The uptake pattern of various crops should be studied prior to fertigation. Nutrient requirement at various stages of crop growth is important.

iii. **Water properties:** Analysis of irrigation water for Ca, Mg, Fe, pH, carbonate and bicarbonate is important for predicting chemical precipitation problems. The risk of precipitation increases with increasing pH or increasing concentration of these materials. Acid may need to be injected with the fertilizer to maintain high nutrient solubility during fertilizer injection.

5.3.8 Fertigation scheduling

Efficient fertigation scheduling requires attention to three factors.

1. Crop and site specific nutrient requirements.
2. Timely delivery of nutrients to meet crop needs.
3. Controlling irrigation to minimize leaching of soluble nutrients below the effective root zone. Nutrients can be injected at various frequencies (daily or bimonthly) depending on systems constraints, soil types and growers preferences. Sandy soils require frequent injections. Since leaching is possible with drip irrigation, nutrients applied in any irrigation must not be subjected to excessive irrigation either during the application or in subsequent irrigations.

5.3.9 Advantages of fertigation

1. Higher water use efficiency has been recorded in almost all the crops tested and hence there is absolutely no doubt that fertigation especially through drip system can economise the irrigation needs of the crops by 30 to 40 percent.
2. The fertilizer use efficiency too has been observed to be markedly improved with fertigation.
3. The better use of fertilizers through fertigation is due to

a. Least loss of nutrients through leaching to around 10 percent as compared to 40 to 55 percent in the traditional system. (the NO_3 –N is found to be accumulated below the root zonal depths and also leached to greater extent with 100 percent recommended fertilizers and furrow irrigation.

b. Optimization of nutrient balance of N : P : K by supplying the desired and required quantities of nutrients directly to the root zone in available forms: since small amounts are provided at regular intervals rather than giving in one or two big doses only, uptake and utilization of nutrients is very high with fertigation. This is largely accomplished due to maintenance of constant concentration of particular nutrient without undue losses through leaching.

c. Fertigation allows for easy application of WSF of different grades and combinations within the rhizosphere so as to match the physiological needs of the crop at different stages of crop growth : For better root development with high P fertilizer initially; for active vegetative growth after establishment with N fertilizers; and for crop maturity, flower and fruit development with high K fertilizers at reproductive stages.

4. Fertigation in almost all cases can effect substantial savings in the application cost fertilizers.
5. Fertigation can be also help in maintaining / improving the physical, chemical and biological conditions of the soil.

5.3.10 Salient findings in fruit crops

Coconut

- Fertigation with WSF at 80% improved trunk girth (6%) and No. of fronds (18%) and fruiting bunches (21.5%) compared to control.
- Yielded more nuts / palm
- Economized 20% fertilizer

Oil palm

- Fertigation with WSF improved trunk girth (18%) and No. of fronds (22%)
- Drip fertigation at 80% WSF yielded 83% more than control
- Saving in fertilizer 20% and water 33%.

Grapes

- Fruit yield with 80% WSF as fertigation was 5770 kg, as against 2635 kg/ha with 100% fertilizer under basin irrigation.

Sapota

- Drip fertigation with 80% WSF – 5800 kg/ha
- Recommended fertilizer + basin irrigation – 4300 kg/ha
- Recommended fertilizer under rainfed – 3080 kg/ha

For an efficient fertigation, the following points must be taken into consideration:

- Fertilizers must be completely water-soluble (e.g. potassium chloride, potassium nitrate, ammonium monophosphate, potassium monophosphate, urea, diammonium phosphate, single super phosphate etc.).
- Water quality should be medium to good.
- If hard water containing high Ca and Mg is available, then use acidic P fertilizers (phosphoric acid, monoammonium phosphate etc.).

- Solution for fertigation should be free from suspended solids and micro-organisms that may plug the small openings in drippers (orifices). To solve this problem, periodic injection of acid in fertigation system is recommended in order to dissolve the plugging material. For this purpose, phosphoric, nitric, sulphuric and hydrochloric acids can be used. Hydrochloric acid is widely used due to its cheapness.
- The irrigation injection system should be completely and carefully washed after the injection of acid.
- When using water with EC more than 2000 milli mhos/cm with a high salinization hazard, we should limit the sodium and chloride containing nitrogen, phosphorus and potassium fertilizers.
- When preparing fertilizer solution for fertigation, some fertilizers must not be mixed together. Such banned mixtures are sodium sulphate with potassium chloride, calcium nitrate and phosphate and sulphate fertilizers, magnesium sulphate with di or mono-ammonium phosphate and phosphoric acid with iron, zinc, copper and manganese sulfate.
- The use of two fertilizer tanks is a better option to solve the above problem. We should place calcium, magnesium and micronutrients in one tank and phosphorus and sulphate fertilizers in the other tank.
- pH should be near about 6 to 7.5; pH more than 8.0 reduces the availability of phosphorus and micronutrients by precipitation. To control the solution pH, use the nitrogen mixture with 80 per cent in nitrate and 2 per cent in ammonium forms.

5.4 Root feeding

Root feeding is one of the methods of fertilize application to crops. The nutrient ions are directly given to the roots in the solution form. This is especially practiced in coconut against several biotic and abiotic stresses. The integrated approaches to tackle these problems include root feeding of nutrients to rejuvenate the affected trees and to impart resistance.

5.4.1 Method of application

Three feet away from the coconut trunk, a pit is dug in search of a fresh root. A freshly developed pencil thickness root of pink or red colour was selected, a slanting cut was given to the root for exposure of vessels. A thick poly bag of 1/2 kg capacity is filled with 200 ml of the solution mentioned below and the cut root is inserted inside the polybag and tied air tightly with a thread. The side roots were cheared to prevent piercing of the poly bag. Below the bag, soil is excavated to seat the bag, so that root is fully immersed in the solution.

Observe the solution for 24 hrs. if the solution is sucked within 4 - 24 hrs, the selection of root is good without any damage. If the solution is not sucked, we have to change the root. The sucking of depends upon sunlight and transpiration rate.

□ □ □

CHAPTER 6

Appendices

APPENDIX – 1: Basic Information

Table 1: Spacing and plant population

Spacing (cm)	Population /acre.	Spacing (Ft).	Population /acre.	Spacing (Ft).	Population /acre.	Spacing (Ft).	Population/ acre.
10 × 10	400000	1' 'X 1'	43560	9' X 9'	537	20' × 20'	108
15 × 10	266866	2' × 1'	21780	10' × 10'	435	21' × 21'	98
20 X 10	2000000	2' × 2'	10890	11' × 11'	360	24' × 24'	75
20 × 15	133333	3' × 2'	7260	12' × 12'	302	25' × 25'	70
30 × 10	133333	3' × 3'	4840	13' × 13'	257	26' × 26'	64
30 × 15	88888	4' × 3'	3630	14' × 14'	222	28' × 28'	55
45 × 15	59259	4' × 4'	2722	15' × 15'	193	30' × 30'	48
15 × 15	177777	5' × 4'	2178	16' × 16'	170	40' × 40'	27
45 × 30	29629	6' × 6'	1210	17' × 17'	150		
60 × 30	22222	7' × 7'	889	18' × 18'	134		
75 × 30	17777	8' × 8'	680	19' × 19'	120		

Table 2: Important tropical fruits and their cultivation details

Fruits Trees

S. No.	Name	Season	Spacing	Pit Size per/ha	No. of plants	Fertilizer dose / tree (in kgs) Compost	N	P	K	Yield / ha (in mt).
1	Mango (*Mangifera irdica*)	Jul-Dec	7 to 10 m (either way)	1 x 1 x 1m	100 to 205	10	0.2	0.2	0.3	8 to 10 (upto 15 years) 15 to 20 afterwards
2.	Acid Lime-(*Citrus aurantifolia*)	Dec-Feb Jun-Sep	5-6 m (either way)	75 x 75 x75cm	275 to 400	10	0.2	0.1	0.1	25 / year
3.	Sweet Orange-(*Citrus sinensis*)	Jul-Sep	7 x 7 m	75 x 75 x75cm	205	10	0.2	0.1	0.1	30
4.	Mandarin Orange-(*Citurs reticulata*)	Nov-Dec	6 x 6m	75 x 75 x75cm	275	10	0.1	0.04	0.05	15-20 years
5.	Grapes- (*Vitis vinifera*)	Jun-Jul	4 x 3 m 3x2m (Muscast)	1 x 1 x 1m	835 1665	50	0.2 0.10	0.08 0.08	0.40 0.30	15–40 year 30 year
6.	Guava-(*Psidium guava*)	Jun-Dec	5 – 6 m (either way)	45 x 45 x 45cm	275 to 400	50	1.0	1.00	1.00	25 year
7.	Sapota-(*Manilkhara achras*)	Jun-Dec	8 x 8m	1 x 1 x 1m	156	50	0.20	0.20	0.30	20 to 25 per year
8.	Papaya –(*Carica papaya*)	Jun-Sep	1.8 x 1.8m	45 x 45 x 45cm	3086	10	0.50	0.50	0.50	100 to 250
9.	Pomegranate (*Punica granatum*)	Jun-Dec	2.5 to 3m (either way)	60 x 60 x 60 cm	1111 to 1600	10	0.20.	0.10	0.40	20 to 25 per year
10.	Jack (*Artocarupus heterophyllus*)	Jun-Dec	8 x 8m	1 x 1 x 1m	156	10	0.15	0.08	0.10	30 to 40
11.	Ber (*Zizyphus mauritiana*)	Jun-Aug	7 x 7m	60 x 60 x 60 cm	205	20	0.20	0.10	0.20	70 –80 kg per tree / year

Table 3: Basic units of measurements

SOME USEFUL DATA

1 Cubic foot of water = 62.4 lbs	1 inch per hour = 1,0083 (approx 1.0) cf/sec./acre
1 Gallon of water weight = 10 lbs	1 litre per second = 30.2 GPM
1Cubic foot per second or cusec = 37.4	1 Kilowatt = 1000 watts = 1.341 HP = 737.5 ft lbs sec.
GPM(GPM = Imperial Gallons per minute)	1 house power = 33,000 ft pounds per minute
1 Cubic foot per second or cusec = 0.9917	1 horse power = 550 ft pounds per second
(approximately 1 acre inch per hour)	1 house power = 746 watts
1 cusec of water for one hour = 100 tons (approx)	1 foot of water exerts a pressure of 62, 425 lb/sq. ft.
1 acre inch water = 101 tons (approx)	1 foot of water exerts a pressure of 0. 4335 lb/sq. Inch.
1 acre foot = 43,560 cubic feet	

Table 4. Conversion Factors

Length		**Area**	
1cm	= 0.3937 inch	1 cent	= 40 sq. m
1m	= 3.2808 ft	1 cent	= 54.5 x 8 feet
1k.m	= 0.62137 mile	100 kuzhl	= 1 maa
1 Inc	= 2.54 cm	3 maa	= 1ac
1 ft	= 0.3048 m	20 maa	= 1vell
1 mile	= 1.60934 km		
Area		**Voulume**	
1 sq.cm	= 0.155 sq.in	1 cu. cm	= 0.061 cu. in
1 sq.cm	= 10.764 sq. ft	1 cu. cm	= 0.035 cu. ft
1 sq.cm	= 0.3861 sq. mile	1 cu. cm	= 0.03519 ft.oz
1 Ha.	= 2.471 ac.	1 litre	= 0.21997 Imp gallon
1 Ha.	= 10,000 sq. m	1 US gallon	= 3.7853 litres
1 ac.	= 43,560 sq. ft	1 cu in	= 16.38 cu. cm
1 ac.	= 100 cents	1 cu ft	= 28.316 cu. cm
1 sq.in.	= 6,4516sq. cm.	1 ft oz	= 28.4131 cu. cm
1 sq.ft.	= 0.0929 sq. m	1 imp gallon	= 4.54596 litre
1 mile	= 2.58999 sq. km	1 imp gallon	= 0.1604 cu. ft
1 ac.	= 0.4047 Ha.	(imp gallon is used in Indian)	
1 ac.	= 4000 sq. ms.		
1 ac.	= 4840 sq. yrds.		
1 sq. mile	= 640 ac.		
1 sq. m.100 centiares	= 1 Centairs		
100 Ares	= 1 Are		
1 cent	= 1 Ha.		
	= 435.6 sq.ft		
Weight			
1 kg	= 2.20462 lbs		
1 lb	= 0.45359 kg		
1 bale (cotton lint)	= 177.8 kgs		
1 million	= 10 lakhs		
1 crore	= 10 millions		

Table 5: Acid equivalent of acid forming fertilizers

Fertilizer	Acid equivalent
Ammonium Chloride	128
Ammonium Phosphate	86
Ammonium Sulphate	110
Ammonium Sulphate nitrate	93
Ammonium nitrate	60
Urea	80

Table 6: Equivalent basicity of basic fertilizers

Fertilizers	Equivalent basicity
Calcium Cyanide	63
Calcium nitrate	27
Dicalcium Phosphate	25
Sodium nitrate	21

Table 7: Optimum pH range for availability of different nutrients

Nutrient	Optimum PH range
N	6.0 –8.0
P	6.5 –7.5
K	6.0 – 7.5
Ca	7.0 – 8.5
Mg	7.0 –8.5
Fe	6.0 and below
Mn	5.0 – 6.5
Bo	5.0 –7.0
Cu	5.0 –7.0
Zn	5.0 –7.0
S	6.0 and above
Mo	7.0 and above

Table 8: Standards for Evaluation water quality from pH and salt an concentration

Total soluble salts (ppm)	Suitable	Doubtful/unsuitable
Up to 400	pH 9.0 <	pH > 9.0
400 – 600	pH 8.5 <	pH > 8.5
600 – 800	pH 8.0 <	pH > 8.0
above 1000	-	unfit for irrigation

Table 9: EC and SAR values for different classes of salinity and sodium hazard of irrigation EC

Salinity class	Class	EC (dsm^{-1})
Low	C_1	0.10 –0.25
Medium	C_2	0.25 – 0.75
High	C_3	0.75 – 2.25
Very high	C_4	2.25 – 5.00

SAR

Sodium hazard	Class	SAR
Low	S_1	10
Medium	S_2	10 – 18
High	S_3	18 – 26
Very high	S_4	26 -31

Table 10: Alkaline soil : Classification

Class of soil	EC (dsm^{-1})	ESP	pH
Saline	>2	-	-
Saline–alkaline	>2	>15	> 8.5
Sodic	-	>15	-

Table 11: Ready reckoner for the preparation of spray fluid

Parts per Million	Percentage	Necessary chemical gm/10 lits.
1	0.0001	0.01
10	0.001	0.10
25	0.0025	0.25
50	0.005	0.50
75	0.0075	0.75
100	0.01	1.00
200	0.02	2.00
300	0.03	3.00
400	0.04	4.00
500	0.05	5.00
600	0.06	6.00
700	0.07	7.00
800	0.08	8.00
900	0.09	9.00
1000	0.010	10.00
1250	0.125	12.50
1500	0.150	15.00
1750	0.175	17.50

Table 12: Ready Reckoner for Dissolving Insecticides at Desired Strength

Desired strength of solution in	Percentage of Active Ingredient of Insecticides in Commercial Products							
%	6.8	18	20	25	30	35	40	46.7
	Quantity of insecticides to be added per litre of water in gram or in ml							
0.005	0.77	0.22	0.25	0.20	0.16	0.14	0.13	0.11
0.010	1.54	0.56	0.50	0.40	0.33	0.29	0.25	0.21
0.020	3.08	1.11	1.00	0.80	0.67	0.57	0.50	0.43
0.025	3.85	1.39	1.22	1.00	0.83	0.71	0.63	0.54
0.030	4.62	1.67	1.50	1.20	1.00	0.86	0.75	0.64
0.040	6.15	1.22	2.00	1.60	1.33	1.14	1.00	0.86
0.050	7.69	2.78	2.50	2.00	1.67	1.43	1.25	1.07
0.060	9.23	3.33	3.00	2.40	2.00	1.71	1.50	1.29
0.080	12.31	4.44	4.00	3.20	2.67	2.29	2.00	1.71
0.1	15.29	5.56	5.00	4.00	3.33	2.86	2.50	2.14
0.20	20.77	11.11	10.00	8.00	6.67	5.71	5.00	4.28
0.25	38.47	13.89	12.50	10.00	8.33	7.15	6.25	5.35
0.3	46.16	16.56	15.00	12.00	10.00	8.57	7.50	6.42
0.4	61.54	22.22	20.00	16.00	13.33	11.43	10.00	8.57
0.5	76.93	27.77	25.00	20.00	66.67	14.29	12.50	10.71
0.75	115.39	41.65	37.50	26.00	24.99	21.44	18.75	14.07
0.8	128.09	44.43	40.00	32.00	26.67	22.86	20.00	17.14
1.0	159.06	55.54	50.00	40.00	33.33	28.52	25.00	21.42

Table 13: Conversion Factors in area

To convert from	To convert to	Multiply (1) by
Ac	Ha	0.405
	Sq m	4046.8
Cu ft.(water)	Cu m	0.028
	1	28.4
Ft	cm	30.48
	m	0.105
Gal	1	4.546

APPENDIX – II: Preparation of Solutions

Preparation of percent

i) Weigh 10 gm of NaCl and ad to it 90 gm of water in beaker. This forms 10% solution NaCl by weight (w/w).

ii) Measure 10 ml glycerine with a pipette and add to it 90 ml of distilled water. This gives 10% solution of glycerine by volume(v/v).

Dilution scheme of a stock solution to desired strengths:

One milligram of any substances dissolved in 1000 ml of solvent is equivalent to one ppm. Similarly 1 g or 1000 mg dissolved in 1000 ml of solvent is equals to 1000 pm i.e. 10 g dissolved in 1000 ml = 10, 000 ppm.

Hence 1 g in 100 ml = 1% = 10,000 ppm

10 g in 100 ml = 10% = 1.00.000 ppm

(100 mg)/ 1 g in 1000 ml = 0.1 % = 1000 ppm

100 mg in 1000 ml = 0.01% = 100 ppm

10 mg in 1000 ml = 0.001 % = 100 ppm

1 mg in 1000 ml = 0.0001% = 1 ppm

A stock solution prepared for 1000 ppm strength can be diluted to the required concentration. Suppose the concentration of the required solution (R) is 25 pp, and the concentration of the stock solution (S) is 1000 ppm concentration, the following table gives some of the ready answers to dilution desired.

Table 1: Preparation of solution of concentration from given stock solution

(S) Stock solution (ppm)	(R) Required solution (ppm)	(S/R) Dilute ml of stock solution to volume (ml)
10,00,000	1,000	1,000
1,00,00	1,000	100
10,000	100	100
1,000	100	10
100	10	10
10	1	10

Equivalent solution

Table 2: Equivalent dilutions (1 gm = 1000 mg = 10,00,000 μg)

%	Gm/1	Mg/1	Ppm/1	Mg/gm
10	100	1,00,000	1,00,000	1,00,000
1	10	10,000	10,000	10,000
0.1	1	1,000	1,000	1,000
0.01	0.1	100	100	100
0.005	0.05	50	50	50
0.001	0.01	10	10	10
0.0001	0.001	1	1	1

To recapitulate, 10 mg of solute diluted to 100 ml with water gives 100 ppm solution. 10 gm of solute diluted to 100 ml water gives 10% solution. One mole of solute + water, volume made upto 1000 ml, gives a molar solution while the same quantity (gm mole) of solute when dissolved in 1000g water gives a molar solution.

Table 3: Strength of aqueous solutions of common acids and ammonium hydroxide

Substances	Specific gravity	Normality	Per ent (W/W)	Milli litres to be taken to make 1 litre of 1 N solution
Hydrochloric acid	1.18	11.3	35	89
Hydrochloric acid	1.16	-	32	98
Nitric acid	1.42	16.0	70	63
Sulphuric acid	1.84	36.0	96	28
Phosphoric acid	1.69	41.1	85	23
Acetic acid	1.05	17.4	99	58
Ammonium hydroxide	0.90	14.3	27	71

APPENDIX – III: Calculaion of Fertilizer Requirement

The nutrients, which are required for plant growth, are available in both organic and inorganic sources. Since the application of organic measures in sufficient quantity is not possible due to its non-availability, the nutrients are to be supplied to the plants only in inorganic form as fertilizers. The plant nutrients as inorganic sources of fertilizers exists in different forms and are expressed as percent viz., Nitrogen as percent N, K as % K_2O, P as % $P_2 O_5$, Mg as % Mg, Ca as % Ca, S as % S, Similarly Cu, Zn, Mo, B, Fe, Mn as %. Hence, it is important to understand the method of calculation of nutrients requirements from the commonly available fertilizers.

Calculation of nutrient content in fertilizers

1. Nitrogen

A. Nitrogenous fertilizers

a) Urea: CO $(NH_2)_2$

Elements	No. of atoms	Atomic weights.	Molecular weights
C	1	12	12
O	1	16	16
N	2	14	28
H	4	1	4
		Total	60

∴ Molecular weight of urea = 60

2 N atoms are present in one molecular weight of urea = 60 gram

Weighty of 2 atoms of N = 14 × 2 = 28 gram

So 60 grams of urea contains = 28 gram

∴ 100 g of urea $= \frac{28 \times 100}{60}$

= 46.6 gram.

Therefore, nitrogen content of pure urea is 46.6 per cent, Depending upon the impurities in the urea, the N content in the fertilizers grade urea weights vary

b) Ammonium sulphate: $(NH_4)_2$ SO_4

Elements	No. of atoms	Atomic weights.	Molecular weights
N	2	14	28
H	8	1	8
S	1	32	32
O	4	16	64
		Total	132 g

Therefore one molecular weight of ammonium sulphate is = 132 g.
2 N atoms are present in 1 molecular weight of ammonium sulphate
The weight of 2N atoms} = 2 × 14 = 28 gram
132 gram $(NH_4)\text{-}_2\ SO_4$ contains = 28 gram N
∴ 100 gram $(NH_4)_2\ SO_4$ contains $= \frac{28 \times 100}{132} = 21.2\%$

N content of $(NH_4)_2\ SO_4$ = 21.2%.

2. Phosphorus

Phosphatic fertilizers

a) Single super phosphate (SSP) $Ca\ (H_2\ PO_4)_2\ H_2O$.

Elements	No. of atoms	Atomic weights.	Molecular weights
Ca	1	40	40
H	4	1	4
P	2	31	62
O	8	16	128
Total			234 g

Molecular weight of SSP = 234 gm
Two P atoms present in one molecular weight of SSP.
Weight of 2P atoms = 2 × 31 = 62 gm.
234 g SSP contains = 62 gram P.
∴ 100 gram SSP. $= \frac{62 \times 100}{234}$
= 26.5%.

Conversion of P (%) to P_2O_5 (%)

1 Mole of P_2O_5 contains = 2 atoms of P.
Molecular weight of P_2O_5

P = 2 × 31 = 62
O = 5 × 16 = 80

142 g

2 P atoms are present in 1 mole of P_2O_5
Weight 2 atoms = 2 × 31 = 62g.
142 gram P_2O_5 contain = 62g P.
62 g P. is present in 142g. P_2O_5
∴ 1g P present in $= \frac{142}{60}$
= 2.29g P_2O_5.

So, 1 g P is present in 2.29 g P_2O_5
Therefore, % P_2O_5 = %P x 2.29
Hence % P_2O_5 present in Ca $(H_2PO_4)_2$ is = 26.5 x 2.29
= 60.7%

So, the chemically pure Mono Calcium Phosphate (MCP)(Ca $(H_2PO_4)_2$ contain 60.7 % P_2O_5.
SSP (Single super phosphate) contains 26.6 % Ca $(H_2PO_4)_2$.
So, 100g SSP 26.6 g MCP 60.7 × 26.6
So, 100 g SSP → 26.6 g MCP →
100
= 16 g P_2O_5

So, SSP contains 16 % P_2O_5.

b) Diammonium phosphate $(NH_4)_2$ HPO_4.

Elements	No. of atoms	Atomic weights.	Molecular weights
N	2	14	28
H	9	13	9
P	1	1	31
O	4	16	64
Total			132

One P atom is present in $(NH_4)_2$ HPO_4. or 132 g $(NH_4)_2$ HPO_4 contains = 31g P.
Therefore, 100 g $(NH_4)_2$ HPO_4 = $\frac{31 \times 100}{132}$
= 23.5 g

Hence, Phosphorus content of pure Deammonium phosphate (DAP) is = 23.5% P.
To convert 23.5% P to P_2O_5.
% P_2O_5 % P × 2.29.
Therefore % P_2O_5 = 23.5 x 2.29 = 53.8 % P_2O_5.

N content of DAP

2 N atoms are present in 1 mole of DAP, or 28 g of N in 132g of DAP
132 g DAP contains 28 g N.
∴ 100 g DAP contains = $\frac{28 \times 100}{132}$
= 21.2% N

So, 100 g DAP contains 21.2 % N and 53.8%. P_2O_5. However, fertilizer grade DAP contains 18% N and 46% P_2O_5 due to impurities.

3. Potassium

A) Potassium fertilizers

a) Muriate of Potash (KCl)

Elements	No. of atoms	Atomic weights.	No. atoms x Atomic weight
K	1	39	1 x 39 = 39
Cl	1	35	1 x 35 = 35
Total		74	

Hence only one K atom is present in KCl

74 gram of KCl contains 39 gm of K.

\ 100 gram KCl contains = $\frac{39 \times 100}{74}$

= 52.7 g or 52.7%

Conversion of % K to % K_2O.

1 Molecular weight of K_2O contains = 2 atoms of K.

∴ Molecular weight of K_2O = 2 x 39 + 16

= 94 g

Weight of 2 atoms of K = 2 x 39 = 78 g

78 g K is present in 94 g K_2O

\ 1 g K is present in $\frac{94}{78}$

= 1.2 g K_2O

So, 1g K ⇔ 1.2 g K_2O

1 g K = 1.2 g K_2O

Therefore % K_2O = % K x 1.2

or

% K = % K_2O ÷ 1.2

Hence % K_2O in KCl = 52.7 × 1.2 = 63.2

94 g K_2O ⇔ 78 g K

1 g K_2O = 0.83 g K

1 g K_2O = 0.83 g K

Hence % K = % K_2O x 0.83.

The purest KCl contains 63.2% K_2O. However, the fertilizer contains around 60% K_2O and this low K_2O in fertilizer grade KCl is due to impurities present in the fertilizers.

b. Potassium sulphate (K_2 SO_4)

Elements	No. of atoms	Atomic weights.	No. atoms x Atomic weight
KSO	2	39	2 x 39 = 78
	1	32	1 x 32 = 32
	4	16	4 x 16 = 64
Total			174 g

K_2SO_4 contain 2 K atoms.
Atomic weight of 2 K atoms = 2 x 39 = 78 gm
172 g K_2SO_4 contains = 78 g K
\ 100 gram K_2 SO_4 contains = $\frac{78 \text{ x } 100}{172}$
= 44.8 g K.

% K_2O = % K x 1.2
∴ = 44.8 x 1.2
% K_2O = 53.7

So, the pure K_2SO_4 contain 53.7 % K_2O. The fertilizer grade contains around 50% K_2O. This reduced K_2O in commercial fertilizer K_2SO_4 is due to the present of impurities.

4. Calcium

a) Gypsum ($CaSO_4$ $2H_2O$)

Elements	No. of atoms	Atomic weights.	No. atoms x Atomic weight
Ca	1	40	40
S	1	32	32
O	6	16	96
H	4	1	4
Total			172 g

172 g $CaSO_4$. $2H_2O$ contains = 40 gm of Ca
∴ 100 gram of gypsum $\frac{40 \text{ x } 100}{172}$ = 23 g Ca

Ca = 23.0%

Conversion of Ca to CaO

Molecular of Cao = 56 gm

40 gm of Ca in 56 gm of Cao

∴ 1 gm of Ca is present in $= \frac{56}{40}$

= 1.4 g

1.4 g Cao = 1g Ca

% Cao = % Cao x 1.4

% Cao = 23 x 1.4 = 32.2%.

5. Magnesium

a) Magnesium sulphate ($MgSO_4 . 7H_2O$)

Elements	No. of atoms	Atomic weights	No. atoms x Atomic weight
Mg	1	24	24
S	1	32	32
O	11	16	176
H	14	1	14
Total			246 g

246 g $MgSO_4.7H_2O$ contains = 24 gm

\ 100 g $MgSO_4.7H_2O$ $= \frac{24 \times 100}{246}$

100 g $MgSO_4 \ 7H_2O$ = 9.75 %

Conversion of Mg to MgO.

40 g MgO = 24 g Mg

1 g Mg = 1.6 g MgO

% MgO = % Mg x 1.6

= 9.75 x 1.6

= 15.6 %

100 g $MgSO_4 . 7H_2O$ contains = 15.6 g MgO.

6. Sulphur

The molecular weight of Ammonium sulphate $(NH_4)_2 SO_4$ is 132 gm. It contain one sulphur atom and its atomic weight is 32 gm

Therefore

132 g $(NH_4)_2 SO_4$ contain = 32g

∴ 100 g $(NH_4)_2 SO_4$ contain $= \frac{32 \times 100}{132}$ = 24 g S

Hence S content of Ammonium Sulphate is = 24%

7. Zinc

a) Zinc sulphate ($ZnSO_{4.}$ $7H_2O$)

Elements	No. of atoms	Atomic weights.	No. atoms x Atomic weight
Zn	1	65	65
S	1	32	32
O	11	16	176
H	14	1	14
Total			287g

287 g $ZnSO_4$.$7H_2O$ contains = 65 Zinc

∴ 100 g of $ZnSO_4$. $7H_2O$ contains = $\frac{65 \times 100}{287}$

100 g of $ZnSO_4$. $7H_2O$ contain = 22.6 g of Zn.

= 22.6% Zn

8. Copper

a) Copper sulphate ($CuSO_4$.$5H_2O$)

Elements	No. of atoms	Atomic weights.	No. atoms x Atomic weight
Cu	1	63.5	63.5
S	1	32	32
O	9	16	144
H	10	1	10
Total			249.5 g

249.5 g Copper sulphate contains = 63.5 g Cu

∴ 100 g Copper sulphate contains = $\frac{63.5 \times 100}{249.5}$ = 25.4 g Cu

% Cu in $CuSO_4$. $5H_2O$ = 25.4 %

9. Iron

a) Ferrous sulphate ($FeSO_4$. $7H_2O$)

Elements	No. of atoms	Atomic weights.	No. atoms x Atomic weight
Fe	1	56	56
S	1	32	32
O	11	16	176
H	14	1	14
Total			278 g

278 g $FeSO_4 . 7H_2O$ contain = 56 g Fe

∴ 100 g $FeSO_4 . 7H_2O$ contain = $\frac{56 \text{ x } 10}{278}$ = 20.1 g Fe

% of Fe present in $FeSO_4 . 7H_2O$ = 20.1%

10. Manganese

a) Manganese sulphate ($MnSO_4 . 4H_2O$)

Elements	No. of atoms	Atomic weights.	No. atoms x Atomic weight
Mn	1	55	55
S	1	32	32
O	8	16	128
H	8	1	8
Total			223 g

223 g. $MnSO_4 . 4H_2O$ contain = 55g Mn

∴ 100 g $MnSO_4 . 4H_2O$ contain = $\frac{55 \text{ x } 100}{223}$ = 24.7 g

% Mn in Manganese Sulphate = 24.7%

11. Boron

a) Borax ($Na_2 B_4O_7 . 10H_2O$)

Elements	No. of atoms	Atomic weights.	No. atoms x Atomic weight
Na	2	23	46
B	4	10.8	43.2
O	17	16	272
H	20	1	20
Total			381.2 g

382 g Boron contains = 43.2 g B.

∴ 100 g Boron contains = $\frac{43.2 \text{ x } 100}{381.2}$ = 11.3 g

% of B in Boron = 11.3%.

12. Molybdenum

a) Ammonium molybdate ($(NH_4)_6$ MO_7 $O_{24}.2H_2O$)

Elements	No. of atoms	Atomic weights.	No. atoms x Atomic weight
N	6	14	84
M	7	96	672
O	26	16	416
H	28	1	28
Total			1200 g

1200 g Ammonium molybdate contains = 672 g of Mo.

\ 100 g Ammonium molybdate contains = $\frac{627 \times 100}{1200}$ = 56 g

% of Mo in Ammonium molybdate = 56%.

Nutrients Supplied by Fertilizers

Nutrient	Fertilizers
1 kg Nitrogen	= 5 kg Ammonium sulphate
	= 4 kg Calcium ammonium nitrate
	= 4 kg Ammonium chloride
	= 2.2 kg Urea
	= 3.3 kg Ammonium nitrate
	= 5.5 kg Di ammonium phosphate
1 kg Phosphorus	= 6.25 kg Super Phosphate
	= 2.02 kg Di ammonium phosphate
1 kg Potassium	= 1.660 kg of Muriate of Potash
	= 2.000 kg of Sulphate of Potash
1 kg Calcium	= 2.5 kg Lime stone
	= 5.0 kg Dolomite
	= 4.5 kg Gypsum
1 kg Magnesium	= 10.4 kg Magnesium Sulphate – $7H_2O$
1 kg S	= 8.3 kg Single super Phosphate
	= 4.2 Kg Ammonium sulphate
	= 5.5 Kg Gypsum
1 kg B	= 8.8 kg Borax = 4.8 Kg SOLUBOR
1 kg Cu	= 4.2 kg Copper sulphate
1 kg Fe	= 5.3 kg Ferrous sulphate
1 kg Mn	= 3.3 kg Manganese sulphate
1 kg Zn	= 4.8 kg Zinc sulphate
1 kg Mo	= 1.92 kg Ammonium molybdate

□ □ □

Colour Plates

Plate 1: Phosphorous Deficiency in Banana

Plate 2: Boron Deficiency in Banana

Plate 3: Potassium Deficiency in Guava

Plate 4: Potassium Deficiency in Mango

Plate 5: Magnesium Deficiency in Grapes

Plate 6: Hen and Chick Disorder in Grapes

Plate 7: Iron Deficiency in Sapota

Plate 8: Iron Deficiency in Acid Lime

Plate 9: Manganese Deficiency in Acid Lime

References

Acharya, C.N. 1939. The hot fermentation process for composting town refuses and other waste materials. Indian J. Agric. Sci., 9:741-817.

Agarwala, S.C. and Sharma, C.P. 1979. Recognising micronutrient disorders of crop plants on the basis of visible symptoms and plant analysis Botany Deptt. Lucknow University, Lucknow.

Alloway, B.J. and Tills, A.R. 1984. Copper deficiency in world crops. Outlook on Agriculture. 13:32-42.

Anonymous. 1987. National symposium on micro nutrient stresses in crop plants. Physiological and genetical approaches to control them. Held at Mahatma Phule Agricultural University, Rahuri.

Anonymous. 1995. Manual on Bio-conversion at Agricultural wastes. Published by Training Division, Directorate of Extension Education, TNAU, Coimbatore. 61 pp.

Anonymous. 2002. National Seminar on Recent trends in sulphur and silicon nutrition of crops. Held between June 12-13 at AC&RI., TNAU, Madurai.

Arvind Kumar, Brijesh Bahadur and Sharma, B.K. 1988. Influence of salts on the germination and seedling growth of barley. Annals of Arid Zone. 27(1):65-66.

Ayers, R.S. and Westcot, D.W. 1976. Water quality for agriculture. FAO, Irrigation and Drainage Paper. 29, Rome. pp.7-11.

Ayers, R.S. and Westcot, D.W. 1985. Irrigation and Drainage 29:1, FAO, Rome.

Balakrishnan, K. and Azhakiamanavalan, R.S. 1999. Calcium for fruit crops. Micronutrient News, XIII 1-3.

Balakrishnan, K. Anbu, S. and Azhakiamanalvan, R.S. 2000. Diagnosis of iron deficiency in fruit crops. Micronutrient News, 14:1-3.

Balakrishnan, K. Anbu, S. and Azhakiamanalvan, R.S. 2000. Potassium and fruit quality. Agro India (Feb.): 20-21.

Balakrishnan, K. Anbu, S. and Azhakiamanavelan, R.S. 1998. Diagnosis of Mn deficiency in fruit crops Micronutrient News, 12:1-4.

Balakrishnan, K. and Azhakiamanavalan, R.S. 1998. Diagnosis of sulphur deficiency in fruit crops. Micronutrient News. 12(12):1-3.

Balakrishnan, K. Rajangam, J. and Thangaraj, T. 2001. Molybdenum. Agro India. (July) 10-12.

Balakrishnan, K. 2000. Chemo and anatomical detection of nutrient deficiency in crops. In: ICAR Summer short course on "Diagnosis and correction of nutritional and physiological disorders in crops" held between June 12-20 at Deptt. of Crop Physiology, TNAU, Coimbatore.

Balakrishnan, K. Anbu, S. and Azhakiamanalvan, R.S. 1999. A key to identify nutritional disorders in horticultural crops and their control measures. In: Souvenir XX Flower, vegetable and fruit show. Held at Govt. of Pondicherry. Deptt. of Agriculture, 62-66.

Balakrishnan, K. Anbu, S. and Azhakiamanavalan, R.S. 1998. Copper deficiency in fruit crops. Micronutrient News, 14: 1-4.

Balakrishnan, K. Venkatesan, K., Thandapani, V. and Sambandamurthi, S. 1994. Diagnosis of zinc deficiency in fruit crops. Micronutrient News. 8: 1-4.

Balakrishnan. K. and Azhakiamanavalan, R.S. 1998. Diagnosis of magnesium deficiency in fruit crops. Micronutrient News, 12(9): 2-3.

Barea, J.M. 1991. Vesicular-arbuscular mycorrhizae as modifiers of soil fertility. Adv. Soil Sci., 15: 1-140

Basak, R.K. 2005. Malnutrition and treatment of vegetable crops. Kalyani publishers New Delhi p. 225.

Bennett, W. 1993. Nutrient deficiencies and toxicities in crop plants. APS Press. Minnesota.

Berger, K.C. 1965. Introductory soils. The Macmillan Company, London.

Bergmann, W. 1992. Nutritional Disorders of plants. Development of visual and analytical diagnosis. Gustav. Fischer Verlag Jena. New York.

Bergmann, W. and Neubert, P. 1976. Plant diagnosis and Plant analysis. Gustav Fischer Verlag Jena, New York.

Bharadwaj, K.K.R. and A.C.Gaur. 1985. Recycling of organic wastes. ICAR Publication, New Delhi. 5 p.

Bhargara, B.S.2002. Leaf analysis for Nutrient diagnosis, recommendation and management in fruit crops. J. Indian Soc. Soil Sci. 50:352-373.

Bhargava, B.S. and Chadha, K.L. 1993. Leaf nutrient guide for fruit crops. In : Ad. Hort. 2: 973-1029 (eds.) K.L. Chadha and O.P. Pareek. Malhotra Publishing House. New Delhi, India.

Bhat, K.L. 2009. Physiological disorders of vegetable crops. Daya Publishing House, New Delhi p.258

Biswas, B.C. and Prasad, N. 1991. Importance of nutrient interaction in crop production. Fert. News. 36: 43-57.

Bose, T.K. Misra, S.K. and Sadhu, M.K. 1988. Mineral nutrition of fruit crops. Naya Prakash, Calcutta.

Bould, C., Hewitt, E.J. and Needham, P. 1983. Diagnosis of mineral disorders in plants. Vol.I. Principles. HMSQ, London.

Bussler, W. 1981. Microscopical possibilities for the diagnosis of trace element stress in plants. J. Plant Nutr. 3: 115-126.

Chapman, H.D. 1984. Suggested fertilizer sampling and handling techniques for determinimg the nutrient status of some field, horticultural and plantation crops. Indian J. Hort., 21: 97-119.

Chapman, H.D. 1966. Diagnostic criteria for plant and soils. University of California, Division of Agricultural Science, Riverside.

Childers, M.F. 1966. Fruit nutrition (2nd). Horticultural Publications. Grainsvile, Florida, USA.

Childers, N.F. 1966. Temperate to tropical fruit nutrition. Horticultural Publications Rutgers. USA.

Chopra, S.C. and Kannan, J.S.1991. Analytical Agricultural chemistry. Kalyani Publishers. New Delhi

Chundawat, B.S. 1997. Nutrient management in fruit crops. Agrotech Publishing House. Udaipur. p. 256.

Clarkson, D.T. and Hanson, J.B. 1980. The mineral nutrition of higher plants. Ann. Rev. Plant Physiol., 31: 239-298.

Cook, R.L. 1962. Soil management for conservation and production. John Wiley and Sons, London.

Das, P.C. 1993. Manures and fertilizers. Kalyani Publishers, Ludhiana.

Devi, G. 1995. Balanced fertilizer use for increasing grain production in southern states. Potash and Phosphate Institute of Canada-India programme, Haryana.

Dilipkumar Das, 1996. Introductory Soil Science. Kalyani Publishers. New Delhi.

Dris, R., Niskanan, R and Jain, S.M. 2001. Crop management and post harvest handling of horticultural products. Vol.I. Quality management. Oxford and IBH Publn. Co. Pvt. Ltd, New Delhi.

Eaton, F.M. 1950. Significance of carbonates in irrigation water. Soil Sci. 69: 123-133.

Epstein, E. 1972. Mineral nutrition of plants. Principles and perspectives. John Wiley and Sons Inc., London.

Ernest, R. and Butlington, I.E. 1976. Crop residues. CBC Handbook of Biosolar Resources. Vol.II.

Fageria, N.K. The use of nutrients in crop plants. CRC Press London p.428.

Fageria, N.K., Baligar,V.C. and Jones,C.A. 1991. Growth and mineral nutrition of field crops. Marcel Dekker, Inc. New York. pp.77-124.

FAO. 1976. Prognosis of salinity and alkalinity. Soils Bull. 31: 268.

Faust, M. 1979. Evolution of fruit nutrition during the 20th century, Hort. Science. 14: 321-325.

Gauch, H.G. 1972. Inorganic plant nutrition Dowden Hutchinson and Ross INC. USA.

Gaur,A.C. 1990. Phosphate solubilizing micro organisms as biofertilizers. Omega Scientific Publishers, New Delhi. 240 p.

Gaur, A.C. and Singh, R. 1982. Integrated nutrient supply system. Fert. News, 27: 87-98.

Gaur, A.C., Neelakantan, S. and Dargan, K.S. 1984. Organic manures. ICAR, New Delhi. Pp.159.

Girdhar, I.K. 1989. Response of sunflower to saline water irrigation and changing root zone salinity. Indian Agric. Sci. 59: 500-503.

Glass, A.D.M. 1989. Plant nutrition: An introduction to current concepts. Jones and Barlett Publishers, Boston M.A.

Goswami, N.N. and Rattan, R.K. 1990. Low analysis fertilizers in Indian agriculture. Fert. News, 35: 15-24.

Goyal, R.S. 1984. Modifications of soil properties due to silicious irrigation water use. J. Indian Soc. Soil Sci. 32: 781-83.

Greenham, D.W.P. 1978. The fertilizer requirement of fruit trees. Proc. Fert. Soc. London.

Grundon, N.J. 1987. Hungry crops: A guide to nutrient deficiencies in field crops. Information Series. G 187002. Queensland Department of Promary Industries. Brisbane, Australia.

Gupta, I.C. 1980. Soil salinity and boron toxicity. Curr. Agric. 4: 1-16

Gupta, I.C. 1987 Predicting sodification in saline water irrigated soils. J. Indian Soc. Soil Sci. 35: 169-170.

Halvin, J.L. Beaton, J.D., Tisdale, S.L., and Neson, W.L. 2004. Soil fertility and fertilizers: An introduction to Nutrient Management . Pearson Education Inc. pp. 86-357.

Hemantranjan, A. and Garg, O.K. 1995. In: Physiological approaches for crop improvement (Ed) Tyagi, D.N. Rajesweri Publication. New Delhi.

Hesse, P.R. 1971. A text book of soil chemical analysis. John Murray, London.p.520.

Ishizuka, Y. 1971. Nutrient deficiencies of crops. Food and fertilizer Technology centre. Taiwan.

Jackson, M.L. 1973. Soil Chemical analysis, Prentice Hall of India Private Limited, New Delhi,p.498.

Jones, R.G.W. and Lunt, C.R. 1967. The function of calcium in plants. Bot. Rev., 33: 407-426.

Jones, S.U. 1979. Fertilizer and soil fertility. Reston publishing company, Virgina.

Jones, J .B. Jr. 1998. Plant nutrition manual. CRC Press, London.

Kabatia, A. and Pendias, H. 1994. Trace elements in soil and plants. 2nd ed. CRC Press, London.

Kalia, H.R. 1964. Anatomical effects of excess and deficiencies of different micronutrient elements on horticultural crops. A review. The Punjab J. Hort., 4: 65-74.

Kamala, T. and Thangaraj, M. 2000. A low cost technique of row feeding of nutrients for coconut. Indian Coconut J., May: 6-7.

Kanear, J.S. and Randhawa, N.S. 1967. Micronutrient Research in soils and plants in India (A Review), ICAR, New Delhi.

Kannaiyan, S. 2000. Integrated nutrient management. TNAU, Coimbatore.

Katyal, J.C. and Agarwala, S.C. 1982. Micronutrients Research in India. Fert. News, Feb.27., 65-86.

Katyal, J.C. and Randhawa, N.S. 1983. Micronutrients. FAO. Rome. Bulletin No.7.

Katyal, J.C. and Sharma, B.D. 1979. Role of micronutrients in crop production. Fertilizer News. (September): pp.33-50.

Kitchen, H.B. (Ed). 1948. Diagnostic techniques for soil and crops. American Potash Institute, Washington D.C., USA.

Kumar, N, Soorianathasundaram, K. and Azhakiamanovalan, R.S. 2002. (Ed). Emerging trends in production technology of tropical fruit crops. ICAR. Summer short course 3-12.7.02, Deptt. Fruits. TNAU, Coimbatore.

Mandal, R.C. 1989. Potassium in field crops and Horticulture. Associated Publishing company, New Delhi.

Marschner, H. 1995. Mineral nutrition of higher plants. 2nd edition. Academic Press. pp.368-379.

Martin-Parvel, P., Gagnard, J. and Gautier, P. (eds.) 1987. Plant analysis: As a guide to the nutrient requirement of temperate and tropical crops. Lavoisier Publishing Co., New York.

Mathen, M.M. 2002. Phosphate solubilizing microorganism in agriculture. Indian Farming, December, 2202. p.35.

Meelu, O.P. and Yadvinder Singh. 1991. Integrated use of fertilizers and organic manures for higher returns. Prog. Fmg., Punjab Agric. Univ., 27: 3-4.

Meena, K.C. 2003. Vermiculture in relation to organic farming. Intensive Agriculture, March – April, 2003. p.27-29.

Meisheri, T.G., Usadedia, U.P., Vaidya, A.C. and Patel, J.B. 2001. Green manures. Indian Farming,(November), 2001. p.11-12.

Mengel, K. and Kirk by, E. 2001. Principles of plant nutrition. Springer International Edition, New Delhi p.849.

Mengel, K. and Kirkby, E.A. 1987. Principles of plant nutrition. Fourth edition. IPI Berne, Switzerland.

Mohan dass, S., Bala Mohan, T.N. and Arjunan, G. 2000. Nutritional Disorders in crop plants. Devi publications, Tamil Nadu p. 188.

Mortvedt, J.J. (ed.) 1991. Micronutrients in agriculture. Second edition. SSSA Book Series No.4. Soil Sci. Soc. Amer., Madison, U.S.A.

Narasimhan, B. 1971. Outlook of nutrition research in mango. The Andhra Agric. J., 18:246-251.

Nason, A. and Mc Elroy, W.D. 1963. Mode of action of the essential mineral elements. In: Plant Physiology. F.C.Steward (ed.) Acad. Press, New York. 3: 436-451.

Negi, S.S. 1998. The Mango. Central Institute of Sub tropical Horticulture, Lucknow.

Nijar, G.S. 1975. Mango nutrition. Punjab Hort. J., 15: 105-113.

Nijar, G.S. 1991. Nutrition of fruits. Kalyani Publishers, New Delhi.

Ohler, J.G. 1979. Cashew communication, No.71. Deptt. of Agrl. Res., Amsterdom.

Olsen, S.R., Cole, C.V., Watanabe, P.S. and Dhan, A.L. 1954. Estimation of available phosphorous in soils by extraction with sodium bicarbonate. U.S.D.A.Circ.939

Palaniappan, S.P. And Annadurai, K. 1999. Organic farming. Theory and practices. Scietific publishers India. Jodhpur.

Pandey, D.P. and Kannan, S. 1979. Absorption and transport of iron in plants as influenced by major nutrient elements. J. Plant Nutr., 1: 55-63.

Piper, C.S. 1966. Soil and Plant analysis, Hans Publishers, Bombay.

Plucknett, D.L. and Sprague. 1989. Detecting mineral nutrient deficiencies in tropical and temperate crops. Series No.7, Westview Press, Boulder Co.

Prasad, R. and Power, J.F. 1991. Crop residue management. Adv. Soil Sci., 15: 205-249.

Rammohan, J., Amina Bibi, K. and Narayanan, A.L. 2002. Recycling of organic waste. Kissan World (Feb), 31-32.

Randhawa, G.S. and Srivastava, K.C. 1986. Citriculture in India. Hindustan Publishing Co. India.

Ranjan kumar Basak. 1999. Fertilizers – A text book. Kalyani Publishers, Ludhiana.

Reuter, D. J. and Robinson, J.B.D. 1997. Plant analysis: An interpretation manual, 2nd edition. CSIRO Publishing Collingwood, Australia.

Rhoades, J.D. 1972, Quality of water for irrigation. Soil Sci. 113: 277-284.

Roy, R.N., Finck, A., Blair, G.J. and Tandon, H.L.S. 2007. Plant nutrition for food security. Discovery publishing House, New Delhi p.348.

Schollenberger, C.J. and Dreibelbis, F.R. 1930. Analytical methods in base exchange investigations in soils. Soil Sci., 30: 161-173.

Sharma, M.K., and Kumar, P. 2011. A guide to identifying and managing nutrient deficiencies in cereal crops. International Plant Nutrient Institute, Georgia, USA p.49

Sharma, M.P. and Adholeya, A. 2001. Mycorrhiza vital for sustainable farming. Agriculture today, December: 44-46.

Sharma, S.K. and Manchanda, H.R. 1989. Effect of irrigation with sodic waters of increasing RSC level on yield and sodium and chloride content of chickpea. J. Indian Soc. Soil Sci. 37:147-151.

Shear, C.B. and Faust, M. 1980. Nutritional ranges in deciduous tree fruits and nuts. Hort. Rev., 2: 142-163.

Shkolnik, M.Y.A. 1984. Trace elements in plants. Developments in crop science (6), Elsevier. New York.

Simpson, K. 1986. Fertilizers and Manures. Longman Publication, London.

Singh, H., Chhonkar, P.K. and Pandey, R.N. 2000. Soil Plant water analysis - a methods manual. IARI, New Delhi. pp.57-59.

Singh, M.V. 2001. Importance of sulphur in balanced fertilizer use in India. Fert. News, 46: 13-18.

Sivanappan, R.K., Krishnamoorthy, K.K. and Manickam, T.S, 1973. Study on quality of irrigation water. Madras Agric. J., 60:785-795.

Smith, P.F. and Scudder, Jr. G.K. 1951. Some studies on mineral deficiency symptoms of mango. Proc. Fla. State. Hort. Soc. 64: 243-248.

Srivastava, A.K. and Singh, S. 2004. Nutrient Diagnosis and Management in citrus. National Research centre for Citrus Nagpur, India p.130.

Stanford, S. and English, L. 1949, Use of flame photometer in rapid soil tests of potassium and calcium. Agron. J., 41: 446-447.

Suarez, D.L. 1981. Relation between pH and SAR and an alternative method of estimating SAR of soil or drainage waters. Soil Sci. Soc. Amer. J. 45:469-475.

Subbiah, B.V. and Asija, C.L. 1956. A rapid procedure for estimation of available nitrogen in soil. Curr. Sci., 25: 259-260.

Subramanian, K.S., Duraisami, V.P. chideswari, T., Govindasamy, M. and Murugappan, V.2004. visual Diagnositic kit for identification and prescription of nutrient disorders in crop plants. TNAU, Coimbatore p.150.

Subramanian, K.S., Kannan, P., Paulraj and Natarajan, S. 2007. Diagnosis of nutrient deficiencies in crops. Department of Soil Science and Agricultural Chemistry, TNAU, Coimbatore, p. 84.

Takkar, P.N. 1996. Micronutrient research and sustainable agricultural productivity in India. J. India Soc. Soil Sci., 44:562-581.

Takker, P.N., Chhibra, I.M. and Mehta, S.K. 1989. Twenty years of Coordinated Research on micronutrients in soils and plants (1967-1987), IISS, Bhopal.

Tandon, H.L.S. 1987. Fertilizer recommendations for Horticultural crops in India. A guide book. FDCO. New Delhi.

Tandon, H.L.S. 1989. Secondary and micronutrient recommendations for soils and crops - A guide book. FDCO. New Delhi.

Tandon, H.L.S. 1993. Methods of analysis of soils, plants waters and fertilizers. (ed). FDCO. New Delhi.

Tandon, H.L.S. 1992. Fertilizers, organic manures, recyclable wastes and biofertilizers. Fertilizer Dev. Corporation. New Delhi.

Tisdale, S.L. Nelson, W.L. and Beaton, J.D. 1985. Soil fertility and fertilizers. Fourth edition. Macmillon Publishing Company, New York.

VanRhee, J.A. 1977. A study of the effects of earthworms on orchard productivity. Pedobiologia, 17:107-114.

Vlek, P.L.G. 1984. Micronutrients in tropical food crop production. Development in Plant and Soil Science. Vol. 14: Junk publisher, Netherlands.

Wallace, T. 1961. The diagnosis of mineral deficiencies in plants by visual symptoms. A colour atlas and guide. His Majesty's Stationery Office, London.

Wittwer, S.H. and Teubner, F.G. 1959. Foliar absorption of mineral nutrients. Ann Rev. Plant physiol. 10: 13-32.

Wolf, B. 1971. The determinations of boron in soil extracts, water and nutrient solutions. Commun. Soil Sci. Plant Anal., 2:363-374.

Yadav, H.D., Dhankar, O.P. and Phugal, V.R. 1985. Effect of chloride and sulphate salinities on germination and growth of cowpea and green gram. Curr. Agric. 9: 55 –58.

Zaidi, PH, Rafiq, S. and Singh, B.B. 1997. Physiology of zinc nutrition in field crops. Fert. News, 42: 63-66.

Ziller, R. 1962. Foliar diagnosis: a method of studying mineral nutrition –its application to the coconut palm. Indian Coconut J., 15:156-159.

□□□